Balloons Aloft:
Flying South Dakota Skies

To the Piedmont Valley Library   August 29, 2013

By Arley Kenneth Fadness

Arley K. Fadness

xulon PRESS

Featuring:

Jules Verne, George Armstrong Custer,
Lieutenant Harlou Neibling, the Stratobowl,
Explorer I and II, Major William Kepner,
Captain Albert Stevens, Captain Orvil
Anderson, Japanese Balloon Bombs, Raven Industries,
Nick Piantanida, Lieutenant
Commanders Malcolm Ross and Lee Lewis,
Radioman 2nd Class Ordien Herr,
Radioman 2nd Class Eugene Herr
and First Lieutenant LeRoy Herr, Skypower,
Susanne and E. Paul Yost,
Sherry and Colonel Joe Kittinger Jr.,
Maxie Anderson, Ben
Abuzzo, Larry Newman, Soukup and Thomas

Table of Contents

A Commemoration

Balloons Aloft: Flying South Dakota Skies commemorates the 150th anniversary of the French novelist, Jules Gabriel Verne's classic, *"Cinq Semaines En Balloon" (Five Weeks in a Balloon)*, written in 1862 and published in 1863.

Jules Gabriel Verne

In 44 chapters, Jules Verne described the imaginary voyage in a double-enveloped balloon – the *Victoria* – across Africa from Senegal to Zanzibar by three daring British aeronauts. Jule's inspiration was ignited after discovering the 1783 balloon papers of Captain Meusnier in the archives of the Academy of Sciences in Paris. Captain Meusnier De La Place, Jean Baptiste Marie-Charles, had conceived a giant dirigible propelled by gas and hot air. In 1862, Jules Verne's friend, Felix Nadar, a French pioneer in photography, described a plan to build a giant aerostat with a two-story wicker house containing double bunks, a kitchen, dining room, and a dark room. The balloon would cost 200,000 francs.

Inspired, Jules Verne began writing about ballooning. One of his first stories to be published described a hair-raising account of the ordeal of an aeronaut who realizes his only companion in the swaying car (gondola) of the balloon is a homicidal maniac.

In 1862, Verne wrote a manuscript dealing with the possibilities of an aeronautical exploration of Africa. After several rejections by publishers, the embittered writer threw the manuscript into the fire. Luckily, his wife, Honorine, retrieved it, and encouraged her husband to send it to one more publisher. He reluctantly agreed. "The publisher, Pierre Jules Hetzel, promptly rejected it. It lacks excitement, adventure was the critique." Discouraged and angry, Jules Verne abandoned the effort with this last publisher. Hetzel, nonetheless, called him back. "Put your documentary into a fictional form." Jules Verne, changed his tune, agreed to comply, and *Cinq Semaines En Balloon* was published in 1863. The book became a smashing hit!

Jules Verne went on to write *Voyages Imaginaires, A Journey to the Center of the Earth (1864), Twenty Leagues Under the Sea, (1870),* and *Around the World in Eighty Days,* (1873). His genius was to use scientific versimiliture, that is, scientific facts to embellish stories.

The rest is literary history. Jules Verne, the third most translated individual author in the world, is affectionately and popularly known as The Father of Science Fiction.

In a life span of 77 years, (born February 8, 1828 – died March 24, 1905), Jules Verne wrote nearly eighty stories on voyages and explorations. He envisioned television, remote control, travel by rocket, the helicopter, the atom bomb, submarines and the dream of going around the world in 80 days.

A later adaption of Jules Verne's *Around the World in Eighty Days,* by S. J. Perelman and made into the movie directed by Michael Todd, inserted part of the journey around the world by balloon. The modern movie actors were David Niven, who played Phileas Fogg, Cantinflas as

Passapartout, Shirley MacLaine as Princess Aouda, and Robert Newton as Inspector Fix. In the adaption of the original novel, Passapartout said, "We are not birds, Master, we cannot fly the mountains. That is not feasible, monsieur, fantastic as it seems...I have made sixty-three ascents, gentlemen, many of them to an elevation of a thousand meters... monsieur...you are addressing the second most celebrated balloonist in Europe." Under Fogg's questioning, Passapartout revealed that the **first** most celebrated balloonist was buried the day before!

While celebrating the 150th anniversary of the writing and publishing of *Cinq Semaines En Balloon*, in 2013, ***Balloons Aloft: Flying South Dakota Skies,*** recalled the 100th anniversary of an actual attempt to accomplish Verne's fantasy. It was in 1962 when Anthony Smith and two cameramen, Alan Root and Douglas Botting, decided to balloon across Africa. The trip turned Verne's fiction into reality. The journey was chronicled in the book *Jambo: African Balloon Safari,* by Anthony Smith. Six flights requiring 24,000 cubic feet of hydrogen, lifted the three aeronauts and sped them across on an amazing safari. The trip was marked by several near disasters. Once they barely avoided a thermal updraft that could have sent the three aeronauts to a dangerous 40,000 feet height. Lions, flash floods and a crash into the side of a mountain were a few of the adventures the trio survived. *"Jambo,"* meant a friendly "hello" in many African dialects. This gentle title belied the challenging and dangerous expedition that ended, thankfully, successfully.

Congratulations, Jules Verne, for 150 years – your *Five Weeks in a Balloon* and so many other fanciful adventures still fascinate young and old alike. Let ***Balloons Aloft: Flying South Dakota Skies*** be a tribute and a thank you to a legendary writer who still inspires today.

A Tribute

"Be not afraid of greatness; some are born great, some achieve greatness, and some have greatness thrust upon them." – Shakespeare, *Twelfth Night*

B*alloons Aloft! Flying South Dakota Skies,* honors two aeronautical legends – the late E. Paul Yost who died in 2007, and Colonel Joesph Kittinger Jr., retired, and very much alive and active today in his eighties.

Ed Yost, Arley Fadness, Joe Kittinger at an
honoring celebration in Bristow, Iowa in 2002

My association with Ed Yost began when in 1978, Ed advertised for a traditional manual drafter for two gas balloon projects. I walked into Ed's corrugated gray steel quonset office/factory at Tea, South Dakota to answer the ad. On the walls were ballooning photos, commemorations and the cover of the 1976 National Geographic Magazine featuring Ed's Atlantic crossing attempt. I introduced myself as "Arley Fadness, interested in the drafting job." I was serving a parish in nearby Harrisburg, Shalom Lutheran Church, but was interested in returning to my earlier profession of drafting, at least, as a hobby in my free time. "What are your qualifications?" Ed barked at me without any pleasantries or idle talk. "Well, I worked for Boeings and..." Ed interrupted me, "You're hired!" That was the shortest interview I have ever had. I had worked for the Boeing Airplane Company in Seattle, Washington many years before. At that time Boeing was working on new, exciting projects such as the 707, the KC135 (which is still in service today), and the Bomarc Guided Missile. I was assigned to the Bomarc Guided Missile department.

Over the years after I left Boeings I worked for Spitznagel Architects, Howard Parezo Architects, and Ken Bastian Engineering, all in Sioux Falls, South Dakota. And later while attending Seminary, I drafted for Gausman and Moore Consulting Engineers and Boonestro and Rosene Consulting Civil Engineers, both in St. Paul, Minnesota. It was clear, Ed Yost was only interested in my Boeing experience. I learned later that he had graduated from the Boeing School of Aeronautics in Oakland, California in 1940. Aha, we were alumni!

I began at once to prepare blueprints for the two gas balloon projects Ed was developing. Both the 1,000 cubic meter and 500 cubic meter helium balloon systems required blueprints for FAA Type Certification.

In 1985, after I moved to North Mankato, Minnesota, I was invited by Ed, to serve as the "chaplain" at the 50th anniversary celebration of Explorer II in the Black Hills, sponsored by The National Geographic Society and Ed Yost.

Then in 2000, my wife Pam and I moved to Custer, South Dakota, enabling me to serve as the Liaison Officer of the newly formed Historical Balloon Society in the Black Hills. Ed was the founder and primary promoter. The purpose of the Historical Balloon Society was to establish a museum and/or interpretive center near the StratoBowl in order to recognize the historic and scientific achievements of Explorer I and Explorer

II as well as Skylab I and IV and other notable ballooning achievements. Ed designed and funded four granite monuments that were placed on the rim of the Stratobowl. The Stratobowl is about 12 miles south of Rapid City, South Dakota. I, Tony Jenniges from Custer, and others, assisted the Bell Monument Company from Kansas in placing the monuments on four concrete slabs.

Presently, a leisure walk in on a Forest Service Trail enables locals and tourists to not only view the scenic natural Bowl, but also read the texts chiseled into the monuments. The texts describe the historic events that took place there in 1934, 1935, 1956, 1959 and in 1960. It was in 1960 that Ed Yost, assisted by Jim Winker and team from Raven Industries, would fly the second successful modern hot air balloon out of the Stratobowl.

When Suzanne Robinson Yost, Ed's wife and balloon partner died in 2001, I presided at the committal service near Alliance, Iowa in 2002. Then, when Ed died in 2007, I presided at his memorial funeral and committal service as well.

It has been a unique privilege to know and befriend Colonel Joseph Kittinger Jr., and wife, Sherry, as fellow collaborators in our several historical endeavors here in the Black Hills of South Dakota. Our mutual efforts, working with Ed Yost, have been to establish an historic landmark as a physical recognition of the Explorer flights out of the Stratobowl. Joe was the celebrated speaker in 1960 at the 25th anniversary of Explorer II. Joe had just completed, and held at that time, the highest parachute jump record from a balloon, at 102,800 feet. The jump took place in New Mexico.

Captain Joe Kittinger, a fighter pilot in the Vietnam war, was shot down and captured. He endured 11 months as a prisoner of war in the infamous "Hanoi Hilton." While a prisoner, he dreamed of flying around the world in a balloon. He was released as a POW, and in 1985 became the first aeronaut to fly across the Atlantic, solo, in an Ed Yost-built-balloon—the *Balloon of Peace*. Prior to this, Joe flew in many gas balloon races, winning enough to even retire one prestigious trophy.

Read the fascinating autobiography of Joe Kittinger in *Come Up and Get Me*, published in 2011. It is co-authored by Joe Kittinger and Craig Ryan with a Foreword by the late moon-walker, Neil Armstrong,

With profound respect, I bow, I salute, I offer this book as a tribute to Joe Kittinger and the late Ed Yost.

Arley K. Fadness
January 1, 2013

Acknowledgments

I am grateful to so many who have given me direction, content, photos and encouragement as we sailed together in this aerial ballooning endeavor. Thanks to Candi Walti, editor of the Aurora County Standard, White Lake Wave and Stickney Argus. Thanks to Jim Winker, Orv Olivier, Ray Summers, Executive director of the Journey Museum, Becky Wigeland, Curator of the National Balloon Museum at Indianola, Iowa, the staff at the Air and Space Museum at Ellsworth Air Force Base, Dan Brosz, Ken Stewart, and Matthew Reitzel, at the South Dakota Historical Society Museum and Archives, Harry Thompson and Elizabeth Thrond at the Center for Western Studies at Augustana, Susan Herr Hanson, and Elaine Herr, Bob Hayes, Helen Wrede, the Custer County Library staff – Jan, Mary, and Doris Ann, Mark McCaffery of Falls Creek Collectibles, Great Falls, Montana, Jesse Sundstrom, Matthew Spencer of the Nebraska Life magazine, Tom Crouch of the Smithsonian Institute, The Thayer County Historical Society and Museum, Siouxland Heritage Museums of Sioux Falls, Stephanie Summerhays of the Smithsonian Institution Scholarly Press, Richard Carlson, Director of the Fassbender Collection at the Adams Museum in Deadwood, Ashley Morton of the National Geographic Society, the Nebraska Prairie Museum, and the staff of Xulon Press. Thanks to my wife Pam, and Paul Horstad.

A special thanks to Norma Kraemer, Reid Riner, Director of the of Minnilusa, Pioneer Museum; and the Raven Industries staff, who proofed my initial manuscript drafts.

Arley K. Fadness

"Arley Fadness has written the definitive book on South Dakota Ballooning. From Smoke balloons at county fairs to the high-flying feats of the Stratobowl flights, this work covers it all.

Readers will especially enjoy learning how ballooning and South Dakota figured in the Second World War. An entertaining and informative read."

Reid Riner, Director
Minnilusa Historical Association
at the Journey Museum, Rapid City, SD

Introduction

Sailing the Paha Sapa

"O Lord, our Lord, how majestic is your name in all the earth." – Psalm 8 (RSV)

We lift off a little after 5:30 a. m. from Ruth Swedland's meadowy green pasture west of Custer. My son Tim and I are guests of Captain Steve Bauer, pilot of one of his six hot air balloons in his Black Hills Balloon fleet. The magic begins as we rise. Up, up and away! The stunning landscape unfolds. Soon we reach a thousand feet. There's no sound except the occasional blast of the propane burner. Hot air spews into the mouth of the multi-colored envelope, like a dragon's hot breath, lifting us ever higher. "Oh my, lookit' that," exclaims one of the other three fellow passengers. This, their first balloon ride, amazes them. "See that, Tim," I murmur as we pass over an old abandoned feldspar mine — rocks strewn and tailings thrust out over the rim.

White-tailed deer look around, but not up. They are startled by the sound and shadow of some alien ghost. "Whosh. Whosh." Captain Bauer gases for another lift. An aroma of lilac and vanilla greets us from a friendly updraft. We pass over Ponderosa Pine forests, then open meadows. We see the dead red pines off in the distance where the pine bark beatles have raised havoc and won. Several evils threaten the peace and tranquility of the sacred Paha Sapa (Lakota for "Black Hills") — beatles, burning, and reckless development.

Floating in the Black Hills
—Photo courtesy of Karen Neal

As we soar, my imagination soars. In my mind's eye, I see Bauer's fantasy balloon float over the city of Custer, named for the famous/infamous Lieutenant Colonel George Armstrong Custer. His legacy is still being debated. We fly east over Custer's encampment where he explored the Paha Sapa in 1874. I see through Paul Horstad's photographer eyes *(Custer Then and Now)*, Custer muster his troops for the long excursion through the Black Hills. There's the spot along French Creek where Horatio Ross and William T. McKay discover gold. Nearby, there's the 7th Cavalry Café (also known as "Wagons West"), Custer Lutheran Fellowship, the Jehovah's Witness meeting place and the Custer State Park entrance.

We wave at the Custer State Park seasonal workers who are unable to sell us a park ticket. Are we trespassing? Below us are herds of wild buffalo, elusive elk, white tail and mule deer, antelope and turkeys. We wonder if the park director will send a helicopter after us to arrest us. We think not. A south breeze pushes us gently past Crazy Horse Mountain. We realize we are in a non-steering vehicle. Now my imagination really takes off. I envision an uplift which enables us to catch a westerly wind towards Mount Rushmore. We see in the distance Orv. Olivier's balloon, *Serenity,* heading for Keystone and the Holy Terror Gold Mine. Then, we too, float right over the top of George Washington's head.

"Serenity"
– Photo compliments of Jr. Wilbert and Jo Vander Woude

Lincoln's nose looks strangely different from this angle. Suddenly we catch a powerful easterly wind that blows us over Hill City. Wind dies down, then reappears from the south and moves us on to the Stratobowl. As we pass over nature's amazing bowl we recall the flights of Explorer I and II in the thirties. This was the beginning of the space age, and it happened here in South Dakota.

We float northward to Rapid City, Lead, Deadwood, Spearfish and land on the top of Devil's Tower. No, not really–my imagination is jolted back to reality when we, in real time and place, pick a landing spot west of Custer. The most obvious is a landing on Highway #16. Champagne and cheese are provided by Captain Bauer. We celebrate, as travelers stop, smile and snap a photo or two.

The Captain prays the balloonist's prayer:

May the Winds welcome you
With softness
May the sun bless you
With his warm hands.
May you fly so high
And so well
That God joins you in laughter
And sets you gently back
Into the loving arms
Of Mother Earth.

Gondola is loaded, envelope wrapped, and we are on our way with memories, real and fanciful, of an unforgettable 2 hour ride in the beautiful Paha Sapas.

Chapter One

A Few Predecessors

"Is is not an infant who has just been born? Perhaps it will be only an imbecile; but perhaps it will become a great man." – Ben Franklin

South Dakota ballooning acknowledges, and is indebted to, its historical predecessors. Of the many adventures and discoveries in history, here are a few examples:

In 7016 B. C., one of two major Sanskrit epics of ancient India – the *Mahabharata*—indicated awareness of aerial transportation. The *Mahabharata*, India's national epic of more than 80,000 couplets or 1.8 million words, ten times the size of the *Iliad* and *Odyssey* combined, is very specific in its indication of the existence of early highly advanced technology that alludes to flight. [1]

Bits of ancient ballooning history suggest that the earliest use of unmanned hot air balloons in the western hemisphere may have been in Peru, about 1500 years ago, during the creation of the Nazca lines. The Nazca lines and figures, or geoglyphs, have been discovered on a high mountain plateau that stretches more than 50 miles. Scientists theorize that hot air balloons were used to sketch out these amazing, mysterious designs. Dick Wirth and Jerry Young in their book *Ballooning* raise the question, "Was the secret of flight known 2,000 years ago?" They go on to say, " In his book *Chariots of the Gods*, Erich von Daniken postulates that the mysterious markings across the Peruvian Nazca Plain were laid out as landing sites by previous visitors to this planet. This is probably the best-known of the many theories on the origins and purpose of these massive designs, which cover over 200 square miles of desert. Made up of long lines of piles of stones, the markings form patterns which can be seen only from the air." .

Researchers backed by the International Explorers Society, agreed with Erich von Daniken, that the designers must have been capable of flight, but on the type of chariot used, their opinion differed. They believed that people of the Nazca civilization, of over 2,000 years ago, had made and flown hot-air balloons, and the International Explorers Society's aim was to prove it.

"The main evidence, a piece of pottery in Lima, bears a design which is unmistakably that of a balloon...using materials which would have been available to the pre-Incan Nazca people, the team based its 1975 design on the simple, one piece tetrahedral shape shown on the pottery. This shape represents the most simple way to fold an uncut piece of material." [2]

Enter South Dakota's interest and efforts in proving or disproving the viability of Erich von Daniken's theory. An envelope was sewn as one spiral gore by Raven Industries of Sioux Falls in South Dakota. Jim Winker, assistant to the designer of the first modern hot air balloon at Raven, Ed Yost, helped on the project in Peru, South America. The airship would be called *Condor I*. The Nazca culture is classified into three ages—the early, middle and late classic 250-750 AD. Geo-glyphs are large drawings in the earth which can only be recognized from the air. There are fish, birds, monkeys, spiders and plants spread on the ground 12 to 800 miles long. As these pictures can be recognized from the air, Jim Winker, and team, demonstrated the possibility of "prehistoric" balloon flight in Peru.

> "The reed gondola, which resembled Thor Heyerdahl's boat *Ra*, was made by Indians at Lake Titticaca. The British balloonist, Julian Nott, was invited to pilot the *Condor I*. His co-pilot was to be Jim Woodman. The crew consisting of more than 20 specialists—Raven technicians, back-up pilots, archaeologists and most importantly, Doc Crane, an American smoke balloonist whose talents were called

on during the inflation...Nott and Woodman, both tall men, rose easily into the sky, sitting astride the reed gondola which had been made to the scale of the diminutive Peruvian Indians. *Condor I* was up to its task, however, and it took the two to a height of about 300 feet. From this position the fliers looked around and realized that they had ascended from the center of a large triangular design which they had not noticed from the ground. The balloon, *Condor I,* had no vents or flying controls, so after a few minutes ballast had to be ejected as the balloon began a fast descent.

On touching down, the two aeronauts stepped out of the gondola and let *Condor I* fly off on its own, The balloon rose handsomely to about 1,700 feet and came down undamaged some 20 minutes later.

The expedition had proved that balloons could have been made and flown in that area in ancient times. The results satisfied Peruvian authorities that the stone lines should be protected monuments." [3]

Hot air balloons were found in Chinese history between 220-280 A.D. There is "one story that tells of an early aeronaut—a woman by the name of T'a Ki, and a court favorite of a Yin dynasty king, Chou-Hsin, (1155-1121 B.C.), who for a wager, prepared a balloon-like vehicle, and made a short but successful ascension." [4]

In the Three Kingdoms Era, Zhuge Liang of the Shu Han Kingdom, invented airborne lanterns called Kong Ming lamps. Whereas flying kites got the Chinese thinking about flight, Zhuge Liang, politician, strategist, diplomat, astrologer and inventor (181-234 A.D.) developed the Kong Ming light, possibly for warfare. It is believed the Kong Ming lantern was an early form of flare for signaling and illuminating the battlefield.

One account reports that when commanding troops were at the front, Zhuge Liang's health began to decline from the pressure. Before his death, Liang invented a light to puzzle the enemy. He fit an oil lamp under a large paper bag causing the bag to float in the air due to the lamp's heating. The enemy was frightened by this light in the air, thinking a divine force was assisting them. Fan Chengda, a poet of the Southern Song Dynasty, writes about Kong Ming's light in a poem: "The candle shoots into the air and stays there." In the Yuan Dynasty, the hot air lantern-balloons became popular especially at festivals.

Albertus Magnus, also known as St. Albert the Great, in the 13th century, suggested a method for constructing an object made of papyrus plant, and filling a container with a mixture of sulfur, willow-carbon and rock salt to produce lift. Though there is no record that these objects actually flew, it is clear that middle age philosophers seriously considered the science and adventure of flying.

The Italian Jesuit, Francesco Lana de Terzi, around 1663, a mathematician, naturalist, and aeronautic, developed a concept for a lighter than air vehicle. Having been a professor of physics and mathematics in Brecia, Lombardy, Lana sketched the concept for a vacuum ship. "He has been known as the 'Father of Aeronautics' for his pioneering efforts, turning the aeronautics field into a science by establishing a theory of aerial navigation by mathematical accuracy." [5]

His airship was never actually built. Francesco de Terzi was conscious that a vehicle such as the one he conceived could be misused as a weapon of war. He wrote, "God will never allow that such a machine be built... because everybody realizes that no city would be safe from raids...iron weights, fireballs and bombs could be hurdled from a great height." [6]

The invention of the hot air balloon is generally and rightly credited to the Montgolfier brothers in 1782...but A. D. Topping, a veritable cornucopia of ballooning surprises and, indeed, the dean of the Akron Lighter-than-Air Society, believes the hot air balloon was invented before that. His source is Jules Duhem's *L'Histoire des Idees Aeronautiques avant Montgolfier*, published in 1943 in Paris.

"According to Duhem and Topping, the discovery was made 73 years earlier, on August 8, 1709. The first successful balloon was demonstrated, they point out, to the King of Portugal by a 23-year-old Jesuit, Bartholomew Laurence de Guzmaon...it consisted of a tray carrying combustibles, covered by a canopy of canvas or heavy paper. According to one story, it was seven or eight feet in diameter. The demonstration took place in a high-ceiling room of the palace in Lisbon.

After Guzmaon had lit the fire, the device rose to a height of perhaps fifteen feet, carrying the fire with it. It drifted against a wall, then fell, setting fire to some hangings.

Guzmaon is best known for his *Passarola*, a widely derided flying machine which Duhem suggests he proposed as a hoax to conceal his true plans. Gibbs-Smith, author of the paper, *Father Guzmaon: The First Practical Pioneer in Aeronautics,* (Journal of the Royal Society of Arts, vol. XCVII), and also his book, *A History of Flying* (London, 1953), takes the *Passarola* more seriously, believing that Guzmaon's original is lost and that only unfriendly caricatures survive.

The extravagant claims made for the *Passarola* by Guzmaon—that it would travel 200 leagues in a day, carry cargo and be used to explore the world from pole to pole—-were apparently taken in all seriousness by his contemporaries. As a result, his actual accomplishment, remarkable as it was, was ridiculed by comparison, and Guzmaon, discouraged, dropped the whole project. Three quarters of a century were to pass before the balloon would come into its own." [7]

The first known living aeronauts would be a sheep, a duck and a rooster.

On June 4th, 1783, the French brothers Joseph-Michel and Jacques-Etienne Montgolfier, flew a hot air balloon with an envelope volume of 28,000 cubic feet, in the marketplace of Annanoy, France.

Several stories, apocryphal or true, preceded this momentous balloon flight. One story told that Joseph Montgolfier was musing on the problem of how the French army could possible storm the British-held Rock of Gibraltar, which was impregnable by land and sea. Joseph, seated in front of the fireplace, noticed his wife's nightgown, hung to dry, suddenly billowed upward to the ceiling by the hot air. Joseph surmised that British forces might be vanquished by an attack from the air. Thus began his quest to devise a vehicle that would fly powered by hot air.

A second story purported that in 1782, Joseph watched the fire in his fireplace as the smoke and sparks rose upward. He then fashioned a small bag of silk and lit a fire under the opening at the bottom causing it to rise. Viola!

A third story stated that Joseph built a 3' x 3' x 4' box out of thin wood and covered the sides and top with taffeta cloth. Under the bottom of the box he crumpled and lit a wad of paper. The box lifted off its stand and promptly collided with the ceiling. Whatever story is true, the Montgolfiers experimented with other lighter than air vehicles until one day in June, 1783, they launched a globe-shaped balloon from that marketplace in Annanoy that rose 6,000 feet and traveled one and a half miles, making the brothers famous overnight.

Later, Etienne, collaborating with Jean-Baptiste Reveillon, a successful wall paper manufacturer, constructed the "Aerostat Reveillon." This balloon with an envelope the size of 37,000 cubic feet carried the first living aeronauts – a sheep, a rooster and a duck. This event was watched by none other than King Louis XVI with his queen Marie Antoinette and a large crowd gathered at Versailles. The flight proved that high altitude air could sustain life. The rooster, sheep and duck landed safely, unharmed except the sheep had kicked one of the rooster's wings and damaged it slightly. The royal palace in the Bois de Boulogne was also the setting for a launch which carried humans in the air for the first time in recorded history. The two daring aeronauts were Pilatre de Rozier and Marquis d'Arlandes.

Soon both hot air and hydrogen gas balloons were invented and/or flown by such notables as J. A. C. Charles, M. N. Robert, James Tyther, Vincenzo Lunardi and Dr. John Jeffries. Frenchman, J. A. C. Charles flew the first unmanned hydrogen balloon, August 27, 1783. Unfortunately the balloon drifted towards the village of Gonesse which frightened the citizenry. They defended themselves against this monster invader from the skies, shooting it with muskets, attacking it with pitchforks, and finally dragging it out of the village by a horse.

Less than two months after the first man-carrying hot air balloon flight, Charles and Marie Noel Robert made a flight of 33 km (20 miles) in the elegant red and gold balloon christened the *Globe*.

September 15, 1784, a minor Italian diplomat, Vincent Lunardi, took off in a 18,200 cubic foot of hydrogen in London. Lunardi flew 13 miles and was hailed a hero. He continued his ballooning career in England until the death of one of his crew in 1786. Unpopular because of the death, he fled England to Europe where he continued flying until his death in 1806.

French Jean-Pierre Blanchard and Dr. John Jefferies (the first American to fly) attempted to cross the English channel but failed. The world's first airman Pilatre de Rozier attempted a channel crossing from Boulogne to England in a hot air/hydrogen hybrid balloon, but he and his partner both crashed and were killed.

It was Jean-Pierre Blanchard who made the first balloon trip to the United States. He was welcomed by General George Washington. Benjamin Franklin, who as ambassador to France, had witnessed the early flights in France, but had died 3 years before the Blanchard flight in America. Blanchard flew, on January 9, 1793, at about 5,000 feet, having been launched from Philadelphia, and landing at Woodbury, New Jersey, some 15 miles away.

Once, someone questioned the worth of flying balloons to Benjamin Franklin, to which he responded with the famous visionary quote, "N'est-il pas en infant qui vient de naitre> peutetre seul un imbecile, main peutetre il deviendra grand homme." (Is it not an infant who has just been born? Perhaps it will be only an imbecile; but perhaps it will become a great man).

American John Wise was an active balloonist in the first half of the 19th century. Wise dreamed and planned for the possibility of crossing the Atlantic Ocean by balloon. "The notion of crossing the Atlantic via balloon as a means of focusing public attention on the potential utility of aeronautics first occurred to Wise during the winter of 1842-1843. The aeronaut had long ago become convinced of the existence of great currents of air in the sky, perpetually blowing in one direction as though they were rivers of air." [8]

In 1905, the Aero Club of nine nations met in Paris and formed the Federation Aeronautique Internationale (FAI), the world's first organization to promote the advancement of flight. Presently, the FAI has over 93 member nations. It regulates all international air sport competition and validates all world records. The FAI oversees balloons, airplanes, parachutists and hang gliders.

By noting a few ballooning predecessors in history, one can begin to understand the allure and adventure by humans for aerial flight.

Chapter Two

The Custer Connection

"My confidence in balloons at the time was not sufficient, however to justify such a course, so I remained seated at the bottom of the basket with a firm hold on either side."
– George Armstrong Custer

An alien creature rose into the gray sky above the Civil War battle-field in 1862. The balloon, a gas powered spy weapon, christened *"The Constitution"* carried the pilot, James Allen, and an unlikely, reluctant observer. Who was this reticent and unlikely observer? Was he the flamboyant Civil War hero they talked about? Was he simply a quiet private plucked from the ranks?

It all began one year earlier, June 18, 1861, when Thaddeus Sobieski Constantine Lowe inflated the balloon *"Enterprise"* at the Columbian Armory on the National Mall where the present National Air and Space Museum now stands. Lowe then ascended, with two telegraph opera-tors in order to demonstrate the feasibility of using balloons to observe enemy movements, and communicate intelligence to the ground. The first airborne telegram was received by President Abraham Lincoln, who quickly gave Lowe the go-ahead to form the first American military balloon corps. In essence, this was the birth of the United States first Air Force!

Later, on September 24, 1861, Lowe ascended to more than 1,000 feet (305 meters) near Arlington, Virginia, across the Potomac River from Washington, D. C., and began telegraphing intelligence on the Confederate troops located at Falls Church, Virginia, more than three miles (4.8) kilometers) away. Union guns were aimed and fired accurately at the Confederate troops without actually being able to see them—a first in the history of warfare.

This triumph led the Secretary of War Simon Cameron to direct Lowe to build four additional balloons. Two more balloons followed shortly after that. The fleet consisted of the *Intrepid, Constitution, United States, Washington, Eagle, Excelsior* and the original *Union*. The balloons ranged in size from 32,000 cubic feet (906 cubic meters) down to 15,000 cubic feet (425 cubic meters). Each had enough cable to climb 5,000 feet (1524 meters).

Lowe ascended scores of times to observe and report on the Confederate Army movements. As a result, the balloon, *"Union,"* became a frequent target. One rebel officer, Eduard Porter Alexander, wrote his father, "we sent a rifle shell so near old Lowe and his balloon that he came down as fast as gravity could bring him." [1]

Lowe wrote later, "a hawk hovering about a chicken yard could not have caused more commotion than did my balloons when they appeared over Yorktown." [2]

A newspaper article in the New York Times published June 2, 1862 described the use of balloons at the battle of Fair Oaks. "Use of the Balloon and Telegraph, Washington," Sunday, June 1: "During the whole of the battle of this morning, Prof. Lowe's balloon was overlooking the terrific scene from an altitude of about two thousand feet. Telegraphic communication from the balloon to Gen. McClellan, and in direct communication with the military wires, was successfully maintained, Mr. Parkspring, of Philadelphia, acting as operator. Every movement of the enemy was obvious and instantly reported. This is believed to be the first time in which a balloon reconnaissance has been successfully made during a battle, and certainly the first time in which a telegraph station has been established in the air to report the movements of the enemy, and the progress of a battle. The advantage to Gen. McClellan must have been immense." [3]

Who then was this reluctant, unlikely observer, ascending upward with Pilot James Allen in Lowe's balloon that particular day? It was the young, ordinarily brash Lieutenant George Armstrong Custer.

George Armstrong Custer

Jay Monaghan in his book, *The Life of General Armstrong Custer* described Custer's initiation into one of T. S. C. Lowe's observation balloons:

> "Next day Union sharpshooters in the rifle pits prevented enemy artillerymen from manning their guns. The Federal fortifications were completed and McClellan's entire line settled down for a siege…at General Baldy Smith's headquarters, a mile from the front, balloonists were sent up to spy behind the enemy line. Smith was dissatisfied with the reports brought down by these aerial professionals. To get more reliable information, he next sent up a military man, General Fitz-John Porter. An unexpected east wind blew this general over the enemy line, and it seemed likely that he would be made a prisoner; however, enemy cannon could not be elevated sufficiently to shoot him down, and a change in the wind floated him back to safety. But the perilous job must now be given to a more expendable man. Why not Lieutenant Custer?

Armstrong was pleased by the recognition, but he was also scared. He had schooled himself to meet death from a falling horse or a flying bullet, but going up in a basket under a fragile bladder of vapor unnerved him. He said later that he tried to appear indifferent when he walked up and climbed in beside Pilot James Allen. The anchor ropes were cast off, and the land sank rapidly. Custer saw many upturned faces, like daisies, watching them. Soon the faces became too small to recognize. Up in space the basket seemed very fragile. He noticed that he could look between the withes and see, far below, treetops and white tents. He asked if the basket was safe for the weight of two men, and his companion frightened him by jumping up and down on the frame to demonstrate its strength." [4]

As the balloon rose into the air, James Allen stood in the basket with his hands on the load ring. Noting that Custer was seated on the bottom of the basket, the aeronaut invited him to stand and enjoy the view. The intrepid cavalryman replied, "my confidence in balloons at the time was not sufficient, however to justify such a course, so I remained seated in the bottom of the basket with a firm hold upon either side." [5]

From Custer's memoirs we get a direct account of his first balloon ascent:

"I first turned my attention to the manner in which the basket had been constructed. To me it seemed fragile, indeed, and not intended to support a tithe of the weight then imposed upon it. The interstices on the sides and bottom seemed immense, and the further we receded from the earth the larger they seemed to become as to whether the basket was actually and certainly safe. He responded affirmatively; at the same time, as if to confirm his assertion, he began jumping up and down to prove the strength of the basket, and no doubt to reassure me. Instead, however, my fears were redoubled, and I expected to see the bottom of the basket giving way, and one or both of us dashed to earth. To the right could be seen the York River, following which the eye could rest on Chesapeake Bay. On the left, and at about the same distance, flowed the James River..." [6].

After a time Custer learned that he must not look down. Instead he must look out at the horizon and battlefield. Using his map as a guide, he located the rivers – York and James — the white tents and general layout. Using his field glasses he could see tiny figures on the road and in the camps.

Eventually he relaxed and began to take notes on what he could see from this amazing height. He writes, *"To the right could be seen the York River, following which the eye could rest on Chesapeake Bay. On the left, and at about the same distance, flowed the James River...Between these two extended a most beautiful landscape, and no less interesting than beautiful; it being made the theatre of operations of armies larger and more formidable than had ever confronted each other on this continent before....I endeavored to locate and recognize the different points of interest as they lay spread out over the surface upon which the eye could rest. The point over which the balloon was held was probably was probably one mile from the nearest point of the enemy's line. In an open country balloons would be invaluable in discovering the enemy's camps, like our own, were generally pitched in the woods to avoid the intense heat of a summer sun; his earthworks along the Warwick were also concealed by growing timber, so that it would have been necessary for the aeronaut to attain the highest possible altitude and then secure a position directly above the country to be examined. With all assistance of a good field glass, and watching opportunities when the balloon was not rendered unsteady by the different currents of air, I was enabled to catch glimpses of canvas through opening in the forest, while camps located in the open space were as plainly visible as those of the Army of the Potomac. Here and there the dim outline of an earthwork could be seen more than half concealed by the trees which had been purposely left standing on their front. Guns could be seen mounted and peering sullenly through the embrasures, while men in considerable numbers were standing in and around the entrenchments, often collected in groups, intently observing the balloon, curious, no doubt, to know the character or value of the information its occupants could derive from their elevated post of observation."* [7]

Custer's initial report must have pleased General Smith. Custer was ordered to fly with the aeronauts almost every day prior to the Confederate evacuation of Yorktown on May 4. Custer made certain claims which are not confirmed by James Allen or Thaddeus Lowe, such as the claim that he invented the idea of flying at dawn and dusk in order to count the Confederate campfires. Custer also claimed credit for the discovery of the evacuation of Yorktown during the night of May 3-4, 1862. However Lowe reported that he had noted nothing unusual during the balloon flight on May 3. [8]

It is not surprising that the gallant, dashing Lieutenant who inscribed on his Toledo sword blade: "Draw me not without provocation, sheath me not without honor" should, as the reluctant balloon observer, quiver abit at his first ascension in a T. S. C. Lowe war balloon.

What then is the South Dakota ballooning connection with George Armstrong Custer? Custer, the Civil War balloonist, left a profound and permanent footprint on South Dakota soil and in South Dakota history. It began when in 1874, Lieutenant Colonel George A. Custer traveled to the Black Hills of South Dakota from Fort Abraham Lincoln near present day Bismarck, North Dakota, as the commander of an exploratory expedition. After months of planning and strategizing, Custer, at age 34, left July 2, 1874 for South Dakota. Assigned to the 7th Cavalry, Custer led 1200 troops, 10 cavalry and 2 infantry companies, an engineer, and artillery detachment, 80 civilians, two miners, Horatio N. Ross and T. McKay, three newspaper correspondents, a botanist, a geologist, N. H. Winchell, Indian scouts, a female colored cook, Sarah Campbell who they called "Aunt Sally", a 16 piece band mounted on white horses playing, "The girl I left behind," as they exited the Fort, the photographer William H. Illingworth, 110 wagons, each pulled by 6 mules, 1,000 horses, 300 head of cattle, 3 gatlin guns, a rodman rifled cannon, several ambulances and a dozen greyhounds.

The expedition lasted 2 months. They marched an average of 18 miles per day. 12 days were spent resting. 880 miles in all were covered over the 60 days.

The 2 expedition miners, Horatio Ross and William McKay, panning for gold along the French Creek, discovered gold within the present city limits of Custer and in larger amounts farther east on the creek. Gold fever erupted! The 1868 Fort Laramie Treaty with the Lakota Indians and the sacredness of the Black Hills was all but ignored as wild- eyed gold seekers flooded the area. In the area where the town of Custer was developing, the population in 1876 exploded to nearly 10,000. A stagecoach line from Cheyenne to Custer began. Four hotels, an estimated 1400 buildings, leantos, tents and other shelters were erected. According to local Custer historian, Jesse Sundstrom, besides the 4 hotels, there were restaurants, a theater in which the first minstrel show was performed in August, dry goods stores, grocery stores, meat shop, a brewery, sawmills and saloons. [9]

On August 10, 1875, the town of Custer was established. As the citizens incorporated, they realized they needed a name for the new city. For awhile the town was called Harney after General William Selby Harney. Eventually, the citizenry chose a new name, voting between two other civil war heroes — General Stonewall Jackson and General George A. Custer. The choice of Stonewall was outvoted in favor of Custer, due to the simple fact that more Union veterans than Confederate veterans lived there. The Custer population, however, declined when gold was discovered up north at Deadwood.

Today, 2013, the city of Custer, population 2,067, might well be considered not only a desirable tourist destination, but also the *"Balloon Capital of the Black Hills."* Interestingly, Custer was the site of the first balloon ascension in the Black Hills in 1882. (See chapter 3). Beyond the fact of George Custer's stint as a balloon observer in the Civil War, the city of Custer is the home of *Black Hills Balloons* which offers Hot Air Balloons rides to locals and tourists, young and old alike. Folks who soar over the Black Hills launched from Custer delight in the serenity and beauty of the sacred Paha Sapa – drifting lazily over canyons, mountains and forests.

Chapter Three

"Smokies" in the Rushmore State

"The Baloon Ascension. According to announcement a grand baloon ascension will take place in Custer on the Fourth. This will be the first has ever occurred in the Hills and will undoubtedly be witnessed by thousands."
– Custer Chronicle, 1882

The earliest recorded balloon flights in the Rushmore State took place during the July Fourth celebration in Custer, South Dakota in 1882. Seven years after the city of Custer was incorporated (1875), the citizenry prepared for this unforgettable event. Three days before the festivities and planned balloon ascension, the Custer Chronicle announced that the event would be, "the first that has ever occurred in the Hills and will undoubtedly be witnessed by thousands of people in the Southern Hills camps."

The Custer Chronicle reported, July 1, 1882: "The Baloon Ascension. According to this announcement, a grand baloon ascension will take place in Custer on the Fourth. This will be the first that has ever occurred in the Hills and will undoubtedly be witnessed by thousands of people in the Southern Hills camps. It was the intention of Mr. Keith and our correspondent to begin their aerial flight about 3 o'clock in the afternoon of the Fourth, but in order to enable people visiting Custer to keep the air-ship in sight for a much longer time, it has been decided to make the start between 7:30 and 8 o'clock p.m. or about sun-down. This sight alone will be well worth a trip to Custer. Mr. Keith is now in Deadwood to purchase the substances necessary for its inflation." [1]

On July 8, 1882 the Custer Chronicle gave the following report: "The day was issued in by the firing of a salute of 39 guns, and from that time until about 9 o'clock the small boy with his firecrackers and the larger ones with their sticks of giant powder were in their element and made the very best use of their time. At 10 o'clock the members of the Sunday School, and others gathered at the Crook street hall and, headed by the brass band, marched to the grove south of town, where swings and picnic dinner interspersed with music by the band were the attractions of the hour. At 2:30 the crowd adjourned to the ball ground where two picked nines under the leadership of B. R. Woods on one side, and W. J. Thornby on the other, occupied the diamond for an hour or so. The game was a hotly contested one, but Wood's nine came out victorious by two tallies the score standing 7 to 9.

The next in order was a race between the firemen's running team with hook and ladder truck and picked men from among our citizens, distance 200 yards. Each team to run and raise a ladder. The firemen were victorious; time 33 seconds. On account of some misunderstanding in the citizens team in regard to raising the ladder the race was lost, as the distance was run in less time than by the other team.

Runaways and indiscriminate equestrian escapades by half intoxicated individuals filled out the remainder of the afternoon until supper time, and resulted in a few skinned faces and peeled elbows, but as it is impossible to kill a drunken man, the riders escaped with only the bruises mentioned above. While we have nothing to say against a drunken individual breaking his own neck, we must denounce the practice as endangering the lives of innocent parties who may be on the streets. On the Fourth while one of these wild horsemen was tearing down the street with his horse on a dead run, R. Wenzel, wife and baby, were crossing and very narrowly escaped being run over, and it was no fault of the rider that they were not. Let the next intoxicated individual who wants to cool his fevered brow on the hurricane deck of a horse, be pulled over by our police officers, and the offender allowed to cool off in the comfortable quarters under the management of Sheriff Code, and it will have a salutary effect. These remarks being brought about the Fourth, we have incorporated them under that head.

Per announcement, at 7:30 o'clock arrangements were made and the work of inflating the monster baloon, was commenced. Under the management of Keith and Benson it was successfully accomplished in about an hour, and at 8:30 the baloon with the bold Aeronauts in the car slowly arose above the housetops. At a height of 80 to 100 feet, as though fully alive as to what was necessary, it created considerable excitement by careening first to one side then to the other, and for a moment it seemed that it would reverse operations and take a trip to the centre of the earth, the next moment, however, it righted and steadily, and gracefully mounted the seriform fluid until it appeared only a mere speck, when it took a northeasterly course and could be traced by the naked eye to some point east of Harney's Peak. The fate of our correspondent and Keith is not yet known, but we are reliably informed that the latter was at Warm Brooks yesterday on his way to Custer." [2]

Though a new aerial phenomenon, the smoke balloon, active in the 1880's, it was conjectured that the Custer balloon may have been inflated with hydrogen, since references suggested that substances, possibly iron and sulphur, were purchased in Deadwood prior to inflation. Iron filings and sulphur mixed together were often used to produce hydrogen. Also, coal gas was another source of inflation for a gas powered balloon.

From 1830 to the 1870's, smoke ballooning grew in popularity, especially at public events. Smoke balloon ascensions became novelty events that featured colorful aeronauts, aerial fireworks and animal and human parachute jumps. In the 70's and 80's this cheaper, quicker method for inflation of the envelope replaced gas. In the past, gas balloons filled with hydrogen, generated by mixing sulfuric acid and iron filings, was the norm. The problem was gas inflation took hours. Launching was tricky. Skill and experience were necessary in maneuvering a gas balloon. The dropping of sand ballast to gain altitude and the venting off of gas to descend or land the basket safely required proficiency. So for fifty years, from about 1880 to 1930, the most common balloon seen in the United States was the smoke balloon. [3]

Smoke balloons were called "smokies." "The Montgolfier brothers were right. 'Phlogiston' can be used to lift people off the ground." Made of muslin or other porous cloth, the balloon envelopes were filled with hot air and smoke from kerosene burners while supported on poles. To inflate his smoke balloons, "Daring Donald MacDonald," the most famous of the smoke balloonists flying at the turn of the century, devised a method of inflation. It consisted of a trench covered with boards. Fire was ignited at one end and the balloon secured at the other end. Once the smoke sealed the cloth and inflated the envelope, the pilot strapped into a harness, the tether is cut and the smokie rises straight up, bucking and swaying at first, then climbing steadily till it reaches around 1,000 feet. In some cases the balloon was held down on the ground by several strong people until the aeronaut signaled them to release. The balloon climbed fast before it cooled. When the balloonist reached the zenith of his/her planned flight, the daredevil jumped and descended by parachute. The balloon turned upside down, emptied through the neck and fell to the earth to be reclaimed and used again. Two chase teams would race to retrieve the balloonist and the balloon itself. "The greatest exponents of the show business at fairs, carnivals and celebrations were W. H. Donaldson and Captain Eddie Allen." [4]

Early beginnings in South Dakota found inventive people experimenting with lighter-than-air aerostats. Toward the end of the nineteenth century, smoke balloons became a common sight at county fairs, city anniversaries and other celebrations.

In 1875, balloons were featured at the Brown County Fair in Aberdeen when Lyman Frank Baum, creator of the Wizard of Oz lived there. On July 4, 1889, the city leaders at Sioux Falls hired Professor Monsieur LeRoy to perform a balloon ascension and parachute jump. During inflation his muslin balloon envelope caught fire and was destroyed in minutes, ending the possibility of a spectacular and entertaining event.

September, 1891, W. A. "Bert" Ward, entertained a gathered crowd at the Sioux Falls Fair. Ward had begun ballooning in Kimball, South Dakota. Ward made two appearances at the fair doing a parachute jump from 1,000 feet. However, later in 1906, Bert Ward's parachute snagged a telephone wire in Monroe, South Dakota, and he fell 70 feet and was killed.

1895-1910 saw the popularity of sport parachuting from smoke balloons increase in interest and attendance at various events. Typical promotion used ads like the one that said, "The Last, The Best" featuring an "Air ship that sails" which appeared in the Sioux Falls Argus Leader, September 18, 1907.

Daring aeronauts flying South Dakota skies included Professor W. Z. Love, Hazel F. Keyes (1896), W. E. Wieterringer (1898), and Fred Butler (September 25, 1895). They employed acrobatics and parachute jumping which always thrilled the crowds.

April 20, 1897, Henry Heintz, a Brookings County businessman and inventor from Elkton, South Dakota, received a patent for his airship design. Heintz conducted an unmanned test flight in 1900, however, after getting 8 feet off the ground, the ship crashed, though the spectators were amazed and delighted. There are no records of further attempts. Henry Heintz, later in 1902, incorporated the Northwest Aerial Navigation Company, but there were no further noteworthy developments.

Balloon inflated at Draper, South Dakota in 1916 - Courtesy of Karen Miller

There are two versions of a balloon ascension celebration at Wall, South Dakota.

The first version taken from the Eastern Pennington County Memories published by The American Legion Auxillary, Carol McDonald Unit, reads: "Balloon ascension at Wall's first celebration on July 10, 1908. A man by the name of Winteringer from Hartington, Nebraska, was to go up in the balloon. He became too intoxicated, and talked to a crippled barber friend from Yankton, S. Dak. into going up in it in his place. The barber knew nothing about a parachute and when the balloon lifted, he became sober enough to become scared. The balloon drifted clear to Sage Creek, before coming down, and it took them until the late hours of the night to get it and the barber brought back to Wall. The barber was unharmed but well sobered up." [5]

A second version, attributed to Ed Yost, speaking at the Old Courthouse museum in Sioux Falls in the 1980's, told this story: "Professor Jack DeElnora was hired to perform a balloon ascension and jump at a 1910 celebration commemorating the founding of the city (Wall, South Dakota). On his way to Wall, DeElnora stopped in Yankton and invited his friend – a disabled barber- to accompany him. Professor DeElnora was an alcoholic, and the residents of Wall soon discovered this problem.

On the day of the balloon flight, the aeronaut completed the necessary preparations for inflating and launching the balloon, and then passed out, intoxicated. Several hundred spectators had arrived to witness the ascension and having waited several hours for it, were determined not to be cheated. The townspeople, subsequently targeted the professor's friend, and the helpless man was dragged to the balloon, placed in the basket and sent aloft. The balloon and its passenger then drifted southward and out of sight. A search party went out the following day and discovered the barber, cold and thirsty, but unhurt, after spending a night with the remnants of DeElnora's aircraft among the jagged peaks of the Badlands." [6]

"Ed Yost, the Father of the modern Hot Air Balloon, said that 'Smoke Jumper' balloons were made of cotton material, then rubbed with soot to make them hold air. According to Ed, the hotel people dreaded having Smoke Jumper balloonists stay in their hotel, as the balloonists would frequently tear up the bed sheets to patch their balloons. These balloons required a large fire. Many early balloonists thought the smoke made the balloon rise, not realizing it was the heat. When the smoke and hot air collected in the balloon, the balloon would rise. The balloonists, who nearly always called themselves 'Professor' would go up in the tethered balloon and then jump out with their parachutes before the amazed audience." [7]

In 1903, a smoke balloon was launched at Deadwood, South Dakota, and caught in a photograph now archived at the Adams Museum. The photo features two female daredevils who were said to have risen 400 feet into the air, and then parachuted safely to the ground. This launch occurred on a lot made vacant in the 1899 fire that destroyed many buildings including the Gem Theater. The Deadwood photo shows a background city of legitimate businesses. The second floor of the businesses, behind the balloon, housed a number of brothels.

In 1903, this smoke balloon ascended over Deadwood, South Dakota. Balloons began making rare appearances in South Dakota starting in Custer in 1882. Balloons were filled with hydrogen or smoke (hot air). Hydrogen was created by mixing sulfuric acid with iron filings. Smoke Balloons were placed over a fire pit forcing heat into the envelope. *– Adams Museum, Deadwood, SD.*

Hazel Keyes was a professional balloonist who usually flew with her pet monkey, Yan-Yan Gin, on her shoulder. She often featured ascending with two trapezes doing acrobatics. Once, December 1, 1894, Hazel ascended with a dog at Yuma, Arizona. Her last recorded ascent was at Sioux Falls in 1896. She made at least 150 ascents though some newspapers reported 200 to 500 in her career.

The most famous female smoke balloonist, Dolly Shepherd, (1887-1983) was not a South Dakotan but an international star. Dolly was called Britain's "Queen of the Air."

Though Dolly used gas balloons, she ordinarily inflated her balloon with hot air by burning a bonfire underneath the envelope. Dolly toured the country, transporting her balloon to community events with four horses and a wagon. When the balloon rose, Dolly hung onto a trapeze below the envelope. When it was high enough she would pull a rip cord and then float down in her parachute.

Dolly Shepherd, Captain Eddie Allen, Peter Kreig and other smoke balloonists were not South Dakotans but are included in this chapter as examples of the scope and popularity of the sport of smoke ballooning.

During the Smoke Balloon era, on June 2, 1908, a free hydrogen balloon, *The Chicago*, landed on the farm of John Draayer, eight miles southeast of Clear Lake, South Dakota. The balloon, *The Chicago,* had been launched from Quincy, Illinois. It drifted over eleven hours. It traveled eight hundred miles, which shattered all speed records up to that time. On board were three passengers – Charles H. Leichliter, Charles A. Coey, and Captain George L. Bumbaugh, plus a fourth, black and tan pup, named Booze. Leichliter, in a newspaper report to the RECORD HERALD writing after the flight, described several incidents during the adventure.

> "None of us had much idea where we were. I have been much over South Dakota and when we first escaped from the clouds I remarked that the country looked like South Dakota or North Dakota, or possible the western portion of Minnesota. After we had landed, Coey and I started for the last farmhouse we had sighted, a matter of a mile away. The master of the farm was just getting the breakfast fire started when we rapped at the door. He was a Hollander. 'What state is this?' was our first greeting. He looked at us

blankly. 'Where are we?' asked Coey, impatient to know. 'Michigan, Iowa, Minnesota, South Dakota?' 'Sout Dakota,' he sang out, understanding us. Then we told him we had just dropped from the clouds. He could not understand at first. Suddenly it dawned on honest John Draayer, the South Dakota farmer.'Oh-h-Oh. You came from opp daire,' and he made a majestic sweep toward the ceiling with his hand. We assured him he was right and from then on he was as much our devoted friend as if we had been angels, dropping in upon him and his family of ten unawares." [8]

Another earlier incident dealt with a doctor from Clear Lake who saw the balloon bump to the ground and then rise up again. The first contact with terra firma was on the farm lot of the Williams ranch. The good doctor raced to town to tell the townsfolk what he saw. They labeled him crazy in the head. He had been involved in an accident two weeks before and the populace concluded he still was not right in the head. So the first place the aeronauts went when they got to town, after they had landed, was to see the editor, C. J. Ronald of the Courier. The editor quickly telephoned the doctor and assured him he was all right and that he in fact had seen the *Chicago* land. [9]

"Smokies" in the Rushmore State and throughout the United States, added much to the history and sport of ballooning. The flamboyant aeronauts brought delight and wonder to thousands.

Chapter Four

Huron-born WW I Aeronaut Hero

"A hero is no braver than an ordinary man, but he is braver five minutes longer." – Ralph Waldo Emerson

"R at a tat tat," the burst of tracers zipped across the sky, clipping two of Lieutenant Paul H. Neibling's parachute shrouds as he floated to the ground. Neibling had just leaped from his balloon-borne wicker basket after he commanded his younger partner, Lieutenant C. Carroll of Garrett, Indiana, to "Go ahead and jump." While dropping and hanging in his parachute, Neibling pulled out his Colt .45 and pumped five shots at the German Fokker D-VII. The Fokker, one of the most feared fighter airplanes in the German arsenal, had just blasted the gas bag above Neibling into flames with its two 7.92 mm spandau LMG 08/15 machine guns. Already three other observation balloons had been shot out of the sky. Was this Fokker pilot a wanna be "Red Baron," the most notable and feared German pilot during World War I? The original "Red Baron" Manfred von Richthofen, accrued eighty combat victories. But no, he was not the "Red Baron" but a determined German pilot. After the burst of shots, Neibling's gun jammed. He struggled with the mechanism, watching and wondering if the Fokker would make another pass at him and finish him off.

"But the enemy plane began to stagger about in the sky. Gradually, the nose went down, and with the engine full on, the plane started a wild, tight spin until at last, to Neibling's amazement, it nosed in at full speed.

The engine went in deep enough to bury the pilot and what was left of the D-VII.

The impact was such, it was impossible to learn what had actually put an end to the Fokker. Nibbling never made a claim for the kill, but to the groundlings, his Colt was the only weapon in the area that challenged the German, and it was admitted generally that in this particular instance, an attacking airplane had been shot done by a balloon observer using only a Colt .45." [1]

Earlier, September 2, 1918, Lieutenant Paul H. Neibling was up alone observing and directing planned barrages on enemy strongholds. Suddenly, two German planes broke through the clouds and quickly blasted the gasbag. Lieutenant Neibling wasted no time. He tumbled over the side of the basket. When he was clear of the basket and his parachute opened, he looked up and saw the balloon was already on fire. He remembered he had a small vest-pocket camera. He took three quick snapshots of the enemy planes as they continued to fire in his direction. Satisfied with that accomplishment, he then decided to use his Colt .45, blazing off several shot at one of the attacking planes. [2]

(Left to right) Captain Birge Clark, First Lieutenant Harlou P. Neibling, Lieutenant William Heftye. 2nd Balloon Squadron. Neibling was promoted to Captain and rated as an Aeronaut, on November 11, 1918. *– Courtesy Mark McCaffrey, Falls Creek Collectibles*

Paul Harlou Neibling was born in Huron, South Dakota, March 12, 1894, eleven years after the city of Huron was incorporated in 1883. Huron, named for the Huron Indians, is located in east central South Dakota. It was settled and developed as a result of the establishment of the Chicago and Northwestern railroad and land boom in the 1880's.

Presently, Huron is the site of the South Dakota State Fair. Several notables regard Huron as their hometown: Cheryl Ladd, one of the original "Charlie's Angels," Gladys Pyle, a congressional legislator and Muriel Humphrey, wife of Vice President Hubert H. Humphrey under President Lyndon Johnson. Less recognized, Lieutenant Paul H. Neibling waits to have significant recognition as a hometown war hero. He died December 9, 1976, in Birch Lake, Minnesota.

As the United States moved closer and closer to the war in Europe, American sport balloonists faced both a challenge and opportunity. Sport Balloonists saw a need to contribute their experiences and skills toward the national defense. An official historian of the U. S. Air Service recognized that, "The United States Army had almost no balloon service previous to our declaration of war." [3]

So with little or no balloon resources in the military, sport balloonists came to the rescue. Fortunately, twelve years earlier in the summer of 1905, the Aero Club of America had organized and launched the sport balloon movement in the United States. "The decision to organize the new club was made by leading members of the Automobile Club of America. These leaders benefited from lectures on aeronautics by Charles Matthews Manly, Samuel Pierpont Langley's chief 'aerodrome' assistant at the Smithsonian Institute. Inspired by Manly's vision of mankind's future in the air, the wealthy automobile enthusiasts funded the Aero Club of America with the vague objective of promoting the 'development of aerial navigation.'" [4]

Various Clubs soon emerged such as the Aero Club of Philadelphia, the Aero Club of New England, the Aero Club of Saint Louis and the Aeronautique Club of Chicago. By 1908, Buffalo, Denver, Baltimore and San Francisco had been added. Specialized balloon clubs at colleges, and the Gordon Bennett races spawned new enthusiasm for sport ballooning. The Gordon Bennett races began in 1906 when James Bennett, an eccentric New York newspaper publisher, established an international balloon prize as an effort to establish his image as a philanthropist of sport.

Sport ballooning thrived prior to World War I but when the guns of 1914 thundered, the Gordon Bennett races ended, since the race was an international annual event.

However, the skills and experiences of Sport balloonists provided the foundation for the development and implementation of U. S. Army Balloon sections. Colonel Charles de Forest Chandler who was closely related to the founding of the Aero Club of America, became the commander of the balloon campaign during the war.

"At the close of World War I, the Balloon Section consisted of 446 officers and 6,365 non commissioned officers and enlisted men. All but 14 of the officers were on flying status. From February 16, 1918, when the 2nd Balloon Company moved into position near Royamieux in the Toul Section, until the end of the war, the officers and men of the Balloon Section made a total of 5,866 ascents in France, spending a total of 6,832 hours in the air. They assisted in 932 artillery adjustments, were attacked by enemy aircraft on 89 occasions, and suffered 35 balloons lost in action to the enemy aircraft and 12 destroyed by German artillery.

American balloon observers were forced to make 116 parachute jumps under enemy fire. They spotted 12,018 enemy shell bursts, reported the presence of 11,856 enemy airplanes and 2,649 German balloons, and observed 400 enemy batteries in action. They reported 22 German infantry movements, 1,113 cases of enemy traffic on roads or railroad, 2,941 smokes, fires, or flares behind enemy lines and 597 explosions." 5

There were a total of 102 American Balloon Companies in World War I. The Balloon Corps had 685 rated balloon pilots and observers. Several sites such as Fort Omaha and Camp John Wise served as training bases.

A typical balloon pilot and team
from the 43rd Balloon Company
—*Courtesy Richard Deschenes Collection*

According to his own report, Neibling started out as a Second Lieutenant with the 1st Minnesota Field Artillery on the Mexican border, but by the time the United States had switched from chasing Pancho Villa to fighting Kaiser Wilhelm, Paul Neibling had reached the rank of First Lieutenant. He finished training at Camp Mills as a member of the renown 42nd (Rainbow) Division.

One morning Neibling's CO marched into a gathering of young officers and asked why no one had volunteered for training in kite-balloon observation. It seemed that the French had made "such an appeal" for American officers to undertake this "interesting" work.

So for no special reason, except that perhaps he was bored with the dreary routine of training, Paul Neibling put his name down and – promptly forgot the incident. A month later he sailed with the 42nd Division aboard the *SS President Lincoln* and subsequently arrived at Coetquidan, France. There was another spell of training that was broken up by an unexpected order to report to the training depot of the French Balloon Corps. The next day he reported to the 73rd Balloon Company located at Maix in the Lune'ville sector. By this time some reservations had set in. Paul was certain there were hundreds of AEF men better fitted for kite ballooning than he, but all that was forgotten during the next week or so when he was taught to plot distances from the air, give instructions to artillery batteries, and keep a balloon log of what was going on about German movements. 6

Both the navy and the army maintained balloon missions. Two types of balloons were used—the free balloon and the captive balloon. The free balloon, pictured as a round ball, was untethered. The captive balloon, looking cigar-like, was tethered and often launched and retrieved by a mobile winch truck. Gondolas were made of rattan, providing resilience and flexibility.

The author examines an original World War I balloon gondola basket displayed at the South Dakota Air and Space Museum at Ellsworth Air Force Base.

Captain Albert Caquot of the French Army adapted the wartime surveillance balloon first developed by the German army. The Caquot balloon, tethered securely, provided a defense in both World War I and World War II as an aerial barrier against attack from the air. However, the captive balloon was a sitting target from enemy fighter places. Captain

Caquot issued his observation crews parachutes so they could jump from their balloons when the bullets began to fly.

The balloonists, such as Neibling, were armed with maps, binoculars, a telephone line to gunners on the ground, and a parachute. The only weapon issued was a holstered hand gun such as the Colt .45 that Neibling carried. Tethered balloons were winched to elevations up to 6,000 feet. Few sights struck fear in the balloonists greater than the approach of hungry enemy planes looking for a kill and spraying their bullets into the highly flammable balloon envelopes. The goal of the enemy was to ignite fire and crash the aerostat.

Lieutenant Glenn Phelps, of the 5th Balloon Company, A. E. F., considered the unofficial American Balloon Ace of the war, had to jump for his life five times in the short space of 4 months from July to November 1918.

"The first of these attacks occurred on July 15, 1918, while Phelps and Lieutenant R. K. Patterson were directing ground fire from an altitude of about 3,500 feet. Artillery rounds began bursting nearby; as they came closer the two men bailed out. Not until they were back on the ground did the two observers learn that the artillery bursts had not been enemy fire after all; they had come from friendly gunners seeking to drive off three enemy aircraft. Half an hour later, Phelps and Patterson were in the air once more.

Then after a long day of relaying information to the ground batteries, the two observers saw five planes droning toward them from their own lines. This time the balloonists assumed they were friendly but discovered otherwise, only when staccato bursts of machine-gun fire sprayed the balloon and its basket with incendiary bullets. Both men promptly dived over the edge of the basket. As Phelps descended, the empty basket plummeted past him with blazing shreds of balloon fabric trailing in its wake.

Three weeks later Phelps was attacked once more, this time by a force of no fewer that 11 German fighters. His balloon enveloped in flames, Phelps made his third jump. In late October he leaped again when his balloon was set upon by a single enemy plane; the balloon did not catch fire, and Phelps was preparing to go up in it again when he discovered that the fabric had been perforated with more than 100 bullet holes. Then on November 10, 1918, came Phelp's most exasperating moment.

His balloon was attacked and burned by two American pilots who had mistaken it for a German observation craft. Once more, he parachuted and landed safely on the ground. It was his last jump. The following day the war was over, and a British Army communique' noted: "All along the Front, the balloons are down." 7

Lieutenant Phelps received the Croix de Guerre and Aero Club of America medal.

Lieutenant Glenn Phelps and South Dakota born Lieutenant Paul Harlou Neibling illustrate the courage and daring typical of audacious aeronauts during the Great War.

Both Phelps and Neibling received the Distinguished Service Cross. Neibling's citation reads as follows: *"The President of the United States of America, authorized by Act of Congress, July 9, 1918, takes pleasure in presenting the Distinguished Service Cross to First Lieutenant (Air Service) Harlou P. Neibling, United States Army Air Service, for repeated acts of extraordinary heroism in action while serving with 2nd Balloon Squadron, U.S. Army Air Service, A. E. F., at Brouville, France, 2 September 1918, and near Fort Du Marre, France, 26 September 1918. While Lieutenant Neibling was making an aerial reconnaissance from a balloon, he was repeatedly attacked by enemy planes, two of which dived at the balloon and opened fire with incendiary bullets. With great coolness he fired at one of them with his pistol and took a picture of the plane with his camera. When the balloon took fire he was forced to jump, but he took two more pictures on the way down in spite of being fired upon. He reascended as soon as a new balloon could be inflated. On September 26, this officer was again attacked while conducting a reglage, but hanging from the basket with one arm, he fired his pistol at one of the enemy planes and jumped only when his balloon burst into flames. He immediately continued his mission in another balloon.*

General Orders: War Department, General Orders No. 46 (1919)
Action Date: 2-Sept-18
Service: Army Air Service
Rank: First Lieutenant
Company: 2nd Balloon Squadron
Division: American Expeditionary Forces" 8

For his bravery, Neibling was promoted to Captain and declared an Official Aeronaut. After the war he moved to St. Paul, Minnesota and worked as an assistant advertising manager for a machinery company. Around 1933, his image appeared on the National Chicle Sky Birds non-sport gum card collector sets.

Chapter Five

The Race for Space begins – Explorer I

"Here the balloon was lightly flying in the face of a light east wind; the sacks of sand which fastened it down had been replaced by twenty sailors." — Jules Verne, *Five Weeks in a Balloon*

"Little did we know that in a few years man would fly in space and then after another few years that man would walk on the moon. But it all started right here in the Stratobowl." – Joe Kittinger

Near Rockerville, halfway between Rapid City and Mount Rushmore in the beautiful Black Hills of South Dakota, lies a national treasure – the Stratobowl. The Stratobowl is a natural circular canyon, a depression in the earth, an amphitheater waiting for the call of history. Then one day, history's day came. The Stratobowl became the launching site of four highly successful, record breaking high altitude balloon flights into the Stratosphere – Explorer I and II, and Stratolab I and IV. Jim Winker, retired engineer from Raven Industries in Sioux Falls, called the Stratobowl the "Cape Canaveral of its day."

Site for Strato-camp setup – *Photo Courtesy of Marlene Dixon, Newcastle, WY*

Explorer I, a stratosphere bound balloon sponsored by the U.S. Army Corps and the National Geographic Society, launched at 5:45 a.m. mountain standard time, on July 28th, 1934. Three courageous army officers, Major William E. Kepner, Captain Albert Stevens and First Lieutenant Orvil A. Anderson, ascended upward into space.

The balloon envelope, fabricated by the Goodyear-Zeppelin Company, filled with hydrogen gas, expanded to 3,000,000 cubic feet in volume, gently lifted the 700 pound, 100 inch diameter gondola with its human and scientific load. The gondola, manufactured by the Dow Chemical Company resembled a giant soccer ball.

Evelyn Chard, a resident of Custer, remembered the historic flights. "I lived in Keystone at the time. My father, Ralph Smith, was the foreman at the CCC (Civilian Conservation Corps) camp in Rockerville. My father built the road into the Stratobowl. We saw all three balloon attempts, the one in 1934, the aborted one, and the one in 1935 that landed successfully near White Lake, South Dakota. My sister Mary Ellen also was there. I remember we went after supper, took our blankets and stayed all night. I remember the bright lights—it was awesome when it went up! And after the launch we went home for breakfast." [1]

Explorer I is ready to launch.
– Photo courtesy of Miller Studio

Evelyn also remembered that there were cabins in the Stratobowl. She spent a week in one of the two cabins during the 1940's with the *"Sub-Deb Club."* She added in a 2012 speech at the Custer Historical Society meeting, "That's the time they came out with a new food–Kraft's macaroni and cheese. Yum Yum!"

In those years, six world powers sparred in the race for supremacy to the stratosphere. Germany, France, Britain, Russia, Belgium and the United States all sought to be first to conquer that rarefied and mysterious region, 8 to 10 miles above the earth.

"November 4, 1927, Captain Hawthorne Charles Gray, from Scott Field at Belleview, Illinois, bid his wife and son good bye and took off into the stratosphere to set a new balloon altitude record. However the next morning, a curious youngster found the balloon in a wood near Sparta, Tennessee. The basket was held fast in the upper branches of one tree, while the envelope was draped over the top of another. The boy called up to the basket, then scrambled up the tree to peer over the edge of the wicker car. A body was curled up in the bottom of the basket. Clad in a fleece-lined flying suit, his face hidden behind an oxygen mask, a parachute strapped on his back, Captain Gray, United States Army Air Corps, had returned to earth. He had flown higher than any other human being in history, but he had not lived to tell the tale." [2] The precious barograph instrument was found indicating that the flight had reached a crest at 42,240 feet.

On May 27, 1931, Auguste Piccard and Charles Kipfer of Augsburg, Germany traveled 9.81 miles or 51,725 feet into the stratosphere by balloon. "Piccard was one of Europe's leading cosmic ray investigators. Knowing that the mysterious radiation originated beyond the earth's atmosphere, Piccard believed it was clear....that the phenomena could best be studied above the atmosphere, which absorbed a portion of the radiation. Piccard considered the possibility of using rockets or an air-plane to lift his instruments into the air, but the choice of the balloon was obvious from the beginning. Able to fly higher and stay longer at altitude, the balloon would also provide a stable instrument platform free from vibration or electrical interference of engines." [3]

September 30, 1933, G. Prokovieff, E. Birnbaum, and K. Godounoff, near Moscow, Russia, ascended 11.6 miles into the air. November 20, 1933, Lieutenant Tex Settle and Major Chester Fordney of Akron, Ohio, lifted to 11.6 miles or 61,237 feet.

Then it was on January 30, 1934, that Paul Fedossyendo, Andrey Vessendo and Iyla Oussyyskine were all killed near Moscow, Russia, having reached 13.67 miles or 72,200 feet. A record holds only if the aeronauts return safely in the basket or capsule.

"In 1933, Army Air Corp's Captain Albert Stevens approached his superiors to launch an expedition into the Stratosphere with a balloon.

Major General Oscar Westover, Army Corps Chief of the Air Corps, recalled his feelings of trepidation when Captain Stevens made the star-tling suggestion that a balloon be constructed approximately forty times times as large as any balloon I had ever flown, and from four to five times as large as any balloon previously flown by man." [4]

Since finances were a challenge, Stevens turned to the National Geographic magazine for funding. The magazine knew Stevens. He had provided photographs for the magazine, one of which was, until 1934, the highest altitude photograph ever taken at 32,220 feet over Dayton, Ohio, in 1924. Society President Gilbert Grosvenor embraced the expedition, and provided most of the financial support.

The Army Air Corps appointed a three-man crew: Major William E. Kepner, pilot; First Lieutenant Orvil A. Anderson, alternate pilot; and Captain Stevens as scientific observer.

While the balloon envelope and gondola were being constructed, Major William E. Kepner and Lieutenant Orvil Anderson searched for an appropriate launch site. Since Kepner served as the Army Project Officer for the flight, he received many suggestions from all over the United States. He traveled to many of those sites without satisfaction. Among the sites considered were Denver; North Platte, Nebraska; the Great Meteor Crater near Winslow, Arizona; Cheyenne, Wyoming; and Salt Lake City, Utah. One place, Lander, Wyoming, appeared promising at first, but after Kepner and Anderson spent the night there, burning old tires filled with oil to study the local winds, they rejected the site.

"The Rapid City Chamber of Commerce was notified when an Associated Press dispatch came into the office of the Rapid City Daily Journal, stating that a location in the northwest was sought. Within 30 minutes, Web Hill, (the father of eye witness, Helen Wrede, 92, presently living in Rapid City) and R. L. Bronson, wired Major Kepner and Captain Stevens, inviting them to inspect sites in the area. This invitation beat one from Denver by an hour and for several weeks Denver and Rapid City were hot rivals for the selection." [5]

Major Kepner received the wire from Rapid City suggesting there was a deep canyon nearby that might be perfect for the project. Kepner and a local cowboy guide rode to the edge of the canyon, Kepner looked down and declared, "God made this site from which to launch balloons." [6]

Immediately the U.S. Army made arrangements to use the private land that has been owned by the same family since the 19th century. This natural depression with a meandering Spring Creek flowing through it was called Moonlit Valley and thereafter, forever, the Stratobowl.

"The bowl is a part of the Bonanza Bar, a placer gold mining firm in operation from about 1890 to 1905, and situated on eighty acres of land in Pennington County. In 1913, John C. Jacobson, an engineer from Wisconsin, visited John Somerud who had been mine superintendent of the Bonanza Bar. Somerud encouraged Jacobson to acquire the stock in the inactive corporation. John Schrader, a Rapid City attorney, was the principle stockholder in the corporation to which the property had been deeded by the United States Government. Somerud died and willed his stock to Jacobson who acquired additional stock and moved onto the property in 1933. Jacobson then granted the National Geographic Society and all others involved permission to use the location for a balloon ascension." [7] Among the present primary owners are Alice (Pat) Jacobson Tomovik and Ken and Cory Tomovik.

In early June of 1934, personnel began arriving at and near the Stratobowl to establish the Stratosphere Flight Camp. The camp was eventually called Stratocamp. Ralph Smith, father of Evelyn Chard and foreman of the CCC, assisted in the development of the new little "city." Lieutenant Orvil Anderson along with the Rapid City Chamber President Web Hill, and South Dakota state officials, directed the establishment of the camp. The Stratocamp consisted of a community of more than a hundred people. Soon the camp had it's own drainage system, sawdust covered streets, parking areas, sewage disposal plant, electric lighting system, waterworks, telephone switchboard, a radio station staffed by NBC technicians and a hospital with an ambulance.

Since hydrogen, a highly inflammable gas, would be used to inflate the envelope, it was necessary to establish a fire department. The fire department included two professional fire fighters, a pump truck, a dozen fire extinguishers and a crew of volunteer fire fighters.

A detail of 120 soldiers from the Fourth Cavalry at Fort Meade, in Sturgis, South Dakota, were assigned duty to the flight project. Immediately, these soldiers with no previous balloon experience began to learn the ropes. They assisted in learning the art of handling large aerostats by launching a 35,000 cubic foot test balloon.

Meteorologists, along with several scientists arrived. They assisted Captain Albert Stevens by preparing and loading the instruments into the gondola. "The actual launch site was 200 feet in a diameter circle in the center of the stratobowl.It was covered with a 4-inch thick layer of sawdust to protect the balloon during inflation." [8]

The stratocamp attracted many visitors, including residents from the nearby Indian reservations, who toured the site wearing traditional Native American garb. "July 9th was a day for ceremony, as the wife of

Governor E.Y. Berry of South Dakota christened the gondola by pouring liquid oxygen over it. It was only a matter of waiting for perfect weather over the 700-800 mile range to the east. Stevens had no intention of going with anything less than ideal weather. 'Photography,' he noted 'was to play an important part in our work during the proposed twelve hours aloft, and for satisfactory photography we must have cloudless skies and good visibility. Only a broad area of high atmospheric pressure could assure such conditions, and we were determined to wait for such a 'high' if it took all summer.'" [9]

Drs. Lyman Briggs and W. F. G. Swann provided professional assistance such as constructing a dehumidifier that would remove excess moisture from the atmosphere inside the gondola.

July 27th the weather appeared right for a flight the next day. A little past dusk the ground crew began inflating the balloon. By two o'clock in the morning the inflation was complete. At that time the envelope held 210,000 cubic feet of highly flammable hydrogen. As the Explorer ascended, the gas would expand and fully inflate the 3,000,000 cubic foot envelope at 65,000 feet. It was estimated that 50,000 spectators had gathered to watch the historic event. Kepner, Anderson and Stevens climbed aboard the gondola and at 5:45 AM Kepner gave the command, "Cast off!" and the mighty airship took off.

Lola Aimenetto, from New Castle, Wyoming, recalled, "my father, John Skog, who was a Swede, was always interested in things scientific, aerial and celestial. He took his children, my two brothers, Russel and Orville, and myself – I was 13 years old- and we drove to the rim of the stratobowl in 1934. The bowl was surprisingly shaped like a bowl and I didn't realize how far down it really went. We had blankets to sit on. Preparations seemed to take forever. There was so much going on. Men were walking around. We lit a fire. My father said to me, 'you can't go down there – it's dangerous.' I saw it go up....it was dark...we couldn't watch for long because of the trees. It seemed to go east (I was on the north side of the bowl). We waited for several hours...then we heard that it crashed. Father came home (we lived in Rapid City) and said it was not too successful. I was not as excited as my father. This is about all I recall." [10]

The three audacious aeronauts busily set about their various duties aboard the black bottomed, white topped spherical gondola. They were the perfect, well trained, experienced officers to attempt such an ambitious, daring excursion into the stratosphere.

Major William Kepner served as the pilot and commander. He was born in Miami, Indiana, January 6, 1893; he son of Mr. and Mrs. Harvey Kepner. At the age of sixteen, he enlisted in the United States Marine Corps. He served until he was honorably discharged after which he studied medicine at the University of Michigan, in Ann Arbor. When war appeared on the horizon, Kepner was commissioned a Second Lieutenant. He was transferred to the infantry, September 11, 1917, and was assigned to the Fourth Infantry, Third Division after which he served overseas. In France Kepner served in the Chateau Thierry sector, and in the battles of Champagne-Marne, Aisne-Marne, Chateau Thierry, St. Amihiel and Mesue-Argonne. Kepner was promoted to Captain on August 5, 1917 and served as a battalion commander throughout the Meuse-Argonne campaign. He was wounded on October 6, 1918, at Madelaine farm. He spent 5 months in the hospital since 1 and 5/8 inches of his jaw bone had been shot away.

Captain Kepner was awarded the distinguished service cross. The citation records that, "while in command of a battalion, Captain Kepner personally led one company of his command in an attack on a woods occupied by a company of German machine gunners. He was the first man to enter the woods and later, when part of the attacking company was held up by flanking machine-gun fire, he, with a patrol of 3 men, encircled this machine-gun, and after a hand to hand fight, put the gun out of action."

In October, 1920, Kepner was sent to the Air Corps Balloon School at Arcadia, California, and then the following May was ordered to Fort Benning, Georgia, to command the Thirty-second balloon company, with which he later took to Lee Hall, Va. Kepner was a twenty-seven year old Captain when he was trained as a Balloon Observer and Dirigible pilot. Kepner flew a gas balloon in the 1928 Gordon Bennett Race and in the National Elimination Balloon Race winning first place for distance covered in a balloon. He finished 3rd and 2nd in similar events at Akron, Ohio, and St. Louis, Missouri. In October, 1930, Kepner was promoted to Major and was transferred to Wright Field as Chief of the Material Division's Lighter-than-air Branch. In 1934, Brigadier General Oscar

Westover assigned Major William E. Kepner as pilot and commanding officer of the Explorer Stratosphere probe.

First Lieutenant Orvil A. Anderson was selected as the second person to fly in Explorer I. He was assigned as the co-pilot and navigator to Major Kepner. Orvil Anderson was born January 10, 1898, in Springville, Utah. Named Orvil Anderson at his birth, an army clerical error changed his name to "Orvil Orson." On August 23, 1917, after he had left Brigham Young University, he enlisted in the Aviation Section of the U. S. Army Signal Corps. When Orvil was young, he had always wanted to be a pilot. He even considered joining the Royal Air Force. Once he enlisted, his requests to be a pilot were either ignored or rejected. What choice did he have left? He decided to go to balloon observer's school. Then on August 3, 1918, he won a commission as Second Lieutenant. Orvil had intended on leaving the Army to study law, however on July1st, 1920, he received a regular commission as a First Lieutenant. In September, 1922, Orvil traveled on the Virginia-to-California first trans-continental airship flight made by a nonrigid C-2, which was a giant dirigible. Three years later, he commanded the 8th Airship Company. Then in 1934, Brigadier General Oscar Westover, Assistant Chief of the Army Air Corps, assigned Orvil Anderson as alternative pilot and operations officer for the Explorer I Mission.

The third flyer for the Stratosphere flight, was Captain Albert Stevens, who was born March 13, 1886 in Belfast, Maine. He graduated from the University of Maine in 1907 with a Bachelor of Science degree, and a Master of Science degree in 1909. Stevens would later marry Ruth Fischer in 1938. Ruth Fischer Stevens would chronicle the life and exploits of Albert Stevens in her book, *My Husband, the First Astronaut,* published in 1987.

Albert Stevens enlisted in the Aviation Section of Signal Corps when World War I broke out. He served with the 88th Aero Squadron, seeing active service in France and Germany. He like Kepner, participated in Chateau Thierry, as well as San Mihiel, and Meuse Argonne Offensives. He worked primarily in photography, taking photos over the Aisne River and over German lines. After the war, on March 21, 1921, Captain Stevens was assigned to the Photographic Unit, Equipment Branch at McCook Field, Fort Bragg, North Carolina. From 1922 to 1932, Stevens served as an aerial photographer, high jump parachute jumper, and photographer for the National Geographic Society. On September 23, 1928, he flew to a height of 37,150 feet in an XCO-5 airplane with pilot Lieutenant Jimmy Doolittle. During the flight, they ran out of oxygen and both Doolittle and Stevens fell semi-conscious for a time, until they reached a lower altitude. Stevens obtained significant aerial pictures from that flight. Later he would fly to 39,150 feet for a photo at peak height near Dayton, Ohio.

Once, Stevens took a photo of President Hoover's residence in California through his Quickwork Photography. He developed the negative in flight in 7 minutes, dropped the photo to a messenger, who transmitted it by telephoto to Washington, and then printed it and placed it on Hoover's desk in 1 hour, 30 minutes, after exposure was made. (Photo 65557) (*My Husband the First Astronaut,* p. 99).

When Captain Stevens proposed a high altitude flight to further scientific knowledge in several fields, the project won backing by the National Geographic Society, the Army, the Army's Chief of Staff, General Douglas MacArthur, as well as the support of President Franklin D. Roosevelt. Thomas McKnew, served as the project's officer.

After the Explorer flights, Albert Stevens would win many awards such as the Distinguished Flying Cross, the Croix de Guerre in France for photographic distinction in World War I under heavy combat, and the Doctor of Science Degree by the **South Dakota School of Mines** on May 30, 1935.

The three aeronauts, Kepner, Anderson and Stevens would collect air samples, photograph the curvature of the earth, measure the intensity of cosmic rays, and gather data regarding the ozone layer.

Soon the airborne laboratory reached 60,000 feet above sea level. Prior to this moment, only eight aeronauts in the world had ever been higher, and of those eight, only five succeeded and lived.

Just then, a clattering noise, a thud, was heard on top of the gondola. The noise caught the flyers attention and concern. The thud had been caused by part of the appendix cord – a small rope- falling on the roof of the gondola. Then there appeared a rip in the bag. Stevens recalled, "Imagine our feelings for a few minutes. It looked as though the valve hose had parted along with the torn fabric. Had this happened we would have been helpless...through the overhead glass porthole we watched the rent in the fabric gradually becoming larger and larger. The minutes passed slowly by; the magnets of the cosmic ray instruments clattered on; the buzzers hammered on the barometer box; the instrument cameras clicked in unison at regular intervals….below us was the brown, sun-baked earth, so far away that no roads, railroads or houses could be made out. Our direction of drift was changing, but that was now a matter of little concern. The question now was not where we should get down, but how!" [11]

They must valve and descend. Soft swishing noises came through the roof which meant a new rent or an increase in length of a rip already there. "Three quarters of an hour passed and we were down to 40,000 feet. Our speed was increasing and a half hour later we were down to 20,000 feet… when suddenly the entire bottom of the bag dropped out." [12]

"Meanwhile, while the drama unfolded high about the now central Nebraska landscape, on the ground "the phone rang at the Gewecke house just as dinner was about over…the caller told everyone (party line) to go out and look up at the sky. Through a brass telescope, the Geweckes stared at a round white ball from which dangled a small black object. Even without a telescope it probably looked closer than it really was. It was about 35 miles west and nearly 11 miles high when they spotted it—somewhere above the Platte River between Gothenburg and Cozad… the Geweckes didn't know what the balloon's crew was just discovering. The Explorer was in trouble, and the lives of everyone aboard were in jeopardy." [13]

Kepner and Anderson cut loose the spectrograph and it floated safely down to earth on it's individual parachute. More ballast was discharged. With parachutes on, it was time to leave. Stevens writes, "the altimeter reading it gave was 5,000 feet above sea level, we were in reality only a little more than half mile from the ground." [14]

Stevens was momentarily blocked from exiting the gondola. He shouted to Anderson, "Hey get your big feet out of the way, I want to jump." Stevens then plunged headlong out of the hatch and soon all three parachutes floated to the ground.

"The gondola hit the ground, with a great thud, landing and crashing in a cornfield on Reuben Johnson's farm near Holdrege, Nebraska. Reuben Johnson had spotted the balloon about the same time that the Geweckes received the general ring on their telephone. Watching the northwest sky, the 35-year-old Swede hitched his team to a cultivator and headed out to his cornfield." [15]

Dorothy Carlson, eye witness to crashing aerostat is holding a piece of
fabric from the envelope. – *Photo courtesy of Nebraska Life Magazine*

"Across the road from the Johnson farm, Dorothy Carlson lay down
to rest while her two-year-old daughter was napping. The 22 year-old had
spent all morning cleaning her big farmhouse top to bottom. Her father-
in-law came in to wake her. 'You'd better come out here and watch this,'
he said. 'It's gonna land pretty soon and it's pretty close.'

By the time Carlson arrived in Johnson's field, the three fliers were
already on the ground, inspecting the smashed gondola and joking with
each other after their narrow escape...the crowd tore down Johnson's
fences, trampled his stunted corn and stampeded his horses and cattle.
They cut off pieces of balloon fabric and picked up broken instruments
as souvenirs." [16]

Stevens later reported, "Major Kepner and I went to the farmhouse
of Mr. Reuben Johnson, on whose field we had landed, to telephone and
send some telegrams. For some time I had been conscious that it was
nearly 100 degrees in the shade...and I still wore two suits of heavy
woolen underwear and a light canvas flying suit; so in the farmhouse I

asked permission to use a room to shed some clothing. In a few minutes I was dressed only in the canvas suit and I took the two suits of underwear outside and hung them over a fence. Then I went inside to get my messages off by telephone. When I came out, I found that souvenir hunters had taken my underwear!" [17]

Crash site near Holdrege, Nebraska
– Courtesy of the Nebraska Praire Museum

The flight appeared to be a failure, since most of the instruments, such as the electroscopes used to gauge cosmic rays, were smashed. The spectrograph was preserved, since it was parachuted to land by it's own individual parachute. Luckily, about 163 photos out of 200 miraculously survived.

All three fliers, Kepner, Anderson and Stevens, received the Distinguished Flying Cross from the Secretary of War. Despite the loss, the flight of what would be known as Explorer I was not a failure. It was a partial success, with the saving of the spectrograph, photos and the achieving of precious flight experience.

With vision, the National Geographic Society President, Grosvenor, had insured the flight with Lloyd's of London. The policy covered the balloon, gondola and scientific instruments to a tune of $30,170.00

The Army and the National Geographic Society began immediately to envision an Explorer II.

Author examines Explorer I crash debris on display at the South
Dakota Air and Space Museum at Ellsworth Air Force Base.
— *Compliments SDAS Museum*

Chapter Six

Reach for Space Success–Explorer II

Balloonist Prayer
"I would like to rise very high, Lord;
Above my City,
Above the World,
Above Time;
I would like to purify my glance
and borrow Your Eyes."
– Michel Quoist

George Moseman was picking corn when he spotted an alien sphere in the southwest sky. Beneath the sphere dangled a round metal ball painted white on top and black on the bottom. The balloon had been launched eight hours earlier from the mountains of South Dakota. The date was November 11, 1935, Veterans Day. Minutes later, George and his brother, John Moseman, would witness the successful landing of Explorer II, practically in the backyard of his Washington township farm near White Lake, South Dakota. [1]

Explorer II, a balloon-gondola flight system built and co-sponsored by the National Geographic Society and the United States Army Air Corps, excited the nation by its 13.71 mile probe into the stratosphere, setting a new altitude record. The pilot, Captain Orvil Anderson and Scientific/ Photographer–Observer, Captain Albert Stevens, redeemed the thrust into the Stratosphere after a near disaster crash in Explorer I. The duo, along with Major William Kepner, had escaped death by a whisker, when they parachuted seconds before a final explosion. Their craft, Explorer I, after soaring over 60,000 feet into the stratosphere, faltered and fell, crashing on farmland near Holdrege, Nebraska. The year was 1934.

The Army Air Corps quickly began to recuperate from the loss of Explorer I by planning a second ascent. The National Geographic Society and the Army Air Corps, together once again, chose the Stratobowl as their launch site. The infrastructure, including the gondola shed, the access road, and a reinforced fence for spectator safety, was still in place from the preparations for Explorer I. The new balloon had an envelope capacity of 3.7 million cubic feet. It was fabricated also by the Goodyear-Zeppelin Corporation, making it the largest balloon constructed at that time. It would not be inflated with volatile hydrogen gas like Explorer I, but helium, to minimize the risks of a gas-induced explosion.

July 12, 1935, the new balloon was ready. Inflation began with thousands of spectators surrounding the site on the rim of the Statobowl. Suddenly, just an hour before the scheduled take-off, there was a loud explosion and the ripped balloon collapsed in seven seconds. Fort Meade soldiers leaped to action and rescued three riggers trapped beneath the fabric, which spread across three acres. A seventeen-foot tear savagely ripped the envelope. The flight would be temporarily aborted.

L. A. "Vern" Kraemer, 92, a pilot since 1940 from Nemo, remembered the aborted flight that July night. He told, "I was dancing with my wife-to-be, as there were dance floors on the east and west ridges of the Stratobowl at the time, when we heard the noise from the rip, as it was about half inflated. Later, when they actually launched, we watched the 1935 flight from our home on Nemo Road, ten miles away by Steamboat Rock." [2]

"The summer of 1935 proved to be very frustrating to the crew and all their associates. It was windy, rainy and of course, cloudy. Hourly reports from the weather bureau were discouraging day after day. The atmosphere in the bowl could be perfect, but winds aloft in an area 700 miles west to 500 miles east was a different story. This was the estimated distance necessary to insure the time needed to complete the flight.

Many of the crew, while waiting for conditions to be just right, learned to trout fish, eat buffalo meat, get reacquainted with the Black Hills, and established long-lasting friendships with local residents. Many were so homesick they had their wives come to join them. Bob Casey, with the <u>Chicago Daily News</u>, said if he was going to 'live' here he was going to register to vote." [3]

"About noon," reported eyewitness, Tom Walsh, "on November 9, 1935, all weather reports came together, and the orders to inflate were given. The patched balloon was spread and the crowds began to gather around the rim of the bowl. After work, the night of November 10, my gang again headed for the scene of action. It was below freezing, so we packed blankets, food and hot coffee, knowing from experience how long it would take to ready this huge balloon for flight.

The Homestake Mine at Lead, South Dakota, had furnished flood lights which encircled the area of activity, so our view during the night was perfect. I might add that now REA serves this rural area. We watched and shivered as the giant grew to several stories high, and the gondola with a ton of instruments and ballast were made fast.

At 7:00 a. m., November 11, the green light was given and the flight began.

The temperature was 5 degrees (above) with a northwest wind at eight miles per hour. Ascent was extremely slow. There were some anxious moments as the balloon began to drift toward the southeast rim of the bowl. Ballast was hurriedly dumped, spewing lead shot over everyone below. But the gondola left the bowl, just barely clearing the trees and hundreds of people on the rim across from us." [4]

Captain Albert Stevens, wrote later, "The balloon envelope, which a few hours before had been a somewhat messy, crumpled

Explorer II on the way to the stratosphere
—Photo courtesy of Fassbender Collection, Adams Museum, Deadwood, SD

wrinkled pile of fabric in a pocket of the Black Hills of South Dakota, was now, in its brief hour of glory, a practically perfect sphere, 192 feet in diameter, and expanded to its full capacity for 3,700,000 cubic feet-the largest sphere, by far, that man had ever constructed for any purpose whatsoever...Our gondola 'living room' was pleasant...on all sides we heard the constant clicking and whirring and buzzing which meant that the many pieces of scientific apparatus were functioning." [5]

Historian Bob Hayes from Keystone
was a launch eyewitness.

Bob Hayes, 85, from Keystone, South Dakota, was one of the eye-witnesses of the launch. Bob was born October 16, 1927 and reported as follows: "I went with my parents when I was 6 or 7 years old. We stayed all night at the rim. Dad drove our 1930 maroon Chevrolet. My Dad was Edward Hayes who worked on the Mount Rushmore monument with Gutzon Borglum. My mother was Gladys Hayes. My mother was also with us at the Stratobowl on Armistice Day and it was cold. Also my Aunt Ellen Hayes was there with her boyfriend, Basil Canfield. Basil was a compressor engineer on Mount Rushmore and he dated Ellen for 20 years and then they got married in 1941.

Back from the rim there were hot dog stands and concession places. There were a lot of people. I went down into the bowl and got into the gondola. There were lots of activities. Prize fighting exhibitions and other entertainments. We saw the aborted one too. It looked like a mushroom and then collapsed to the ground. When the balloon went up, the pilots were riding on top of the gondola and then they released the lead ballast when it stalled which gave the needed lift. I remember leaving the Stratobowl one of the times. I was with John Twining, my grandfather, John Martin Hayes, and my dad. We had parked between the rim and the entrance. There was so much traffic, John Twining got out and stopped the traffic so we could get out." [6]

Next, meet another eyewitness, Helen Wrede, 92, interviewed in her home in Rapid City on April 21, 2011.

Helen was the daughter of Web Hill, who was President of the Rapid City Chamber of Commerce at that time and deeply involved in the hospitality role for the National Geographic team, the scientists and the fliers. Helen, at age 16, saw both Explorer I and Explorer II launch from the bowl. "It was cold as we sat in the car and watched the inflation. It was awesome! We were standing when Explorer II lifted off and my friend from Buffalo Gap stood next to me and squeezed my arm when the balloon stalled. Then we heard Captain Albert Stevens hollered, 'Blast, Andy, Blast!'

Later, my dad, Web, mother Ruth, and brother John and I traveled to Washington, D. C. and were honored to stay with the Secretary of the National Geographic Society, Dr. Tom Mcknew." [7]

Eyewitness Helen Wrede of Rapid City
– Photo courtesy Minnilusa Historical Society

The Explorer II flight would reach a record altitude of 72,395 feet or 13.71 miles about sea level. That computes to 96/100ths of the mass of the atmosphere.

"Captain Stevens designed the life support, or 'air conditioning' system. He used a mixture of 46% liquid oxygen and 54% liquid nitrogen to replenish the cabin atmosphere. Among the many 64 scientific experiments tested in the stratosphere were cosmic ray, sunlight brightness, photography, spores, 12 gallons of Stratosphere air, fungi and ozone calculations.... Medical scientists also used Explorer II to see what effect cosmic radiation might have on living organisms by sending Drosophila fruit flies aloft. Although they were carried inside the gondola, all of the adult flies in the package died due to the cold temperatures encountered during the flight. At first, the scientists feared the experiment would be a total loss; then they found some of the larvae and eggs survived, and they ended up with 98 individuals for breeding. Subsequent results were inconclusive for any radiation-induced mutations due to the small number of flies returned." [8]

"Throughout the flight, Stevens talked with engineers from the National Broadcasting System (NBC) who were on the ground at Rapid City....a little past 2 p.m., the aeronauts talked directly to the Pan American Airways, 'China Clipper,' which was flying over the Pacific Ocean en route from San Diego to San Francisco." [9]

The Explorer II flight was an astounding success.

Another eyewitness of the launch was Clara Kieffer. When interviewed by the Stickney Argus in 2010, she reported, "I was at the Stratobowl in Rapid City on the day the Stratosphere Balloon was launched. My future husband, Lawrence Kieffer was at the landing site later that day. We had never met and had no idea that we would one day be married." [10]

So how and where would the Explorer II land? Would it land safely? At National Geographic headquarters in Washington, D. C., officers and staff studied a map of South Dakota and surrounding states. Someone suggested an office pool of sorts, with participants guessing near which town Explorer II might land. Elsie Grosvenor, wife of National Geographic Society president Gilbert Grosvenor, said she put her money on Stickney, South Dakota.

"When the balloon touched down, softly, eight hours and 13 minutes after take-off on a section of school land, 12 miles south of White Lake and 15 miles southeast of Kimball, that was close enough to Stickney for Elsie Grosvenor to pocket $14, back in Washington, D. C." [11]

Clara Kieffer tells her story at Explorer II's 75th anniversary celebration at White Lake. – *Photo courtesy of Candi Walti*

George Moseman recalled, "in those days we had no newspaper and weren't too well informed. I heard motors running. Only motors, I thought, around, were the county gravel trucks. But it was cars coming from everywhere. It was in the middle of the afternoon about 3:00 and my brother John and I walked out of the field. I said to John, 'it's going to come northwest—close to us.' So we went over to John's place. John's son went to the nearby Christian Day School, and the minister, I remember, dismissed school, piled the kids in his car, and headed for the balloon. We took a single-horse buggy and went one mile south to the Just farm, tied the horse up and walked west to where the balloon was landing. We saw Captain Albert Stevens and Captain Orvil Anderson who piloted the balloon and watched them pull the rip cord." [12]

Other witnesses of the landing near White Lake were Caryl Schone, Norma Steffen, Marvin Steffen, Peggy Wright, Don Reeves, Lowell "Bud" Smith, Goldie Messer Evans and Lou Ora Nelson Busk Houk. Lowell Smith told the Stickney Argus, "I was eight years old and attending school two miles west and three miles south of Iona, South Dakota. One boy asked the teacher if he could to the bathroom (in the outhouse outside the school) and he came back into the schoolhouse to tell the teacher, 'Something is going over the school!' There were 28 students in that one room school, and the teacher dismissed all of us to go outside and look. The balloon was 4,000 feet up at that point. It was something you don't forget." [13]

Robert Plut, a retired pharmicist from Seattle, Washington, recalled the following eyewitness experience in a telephone interview with this author.

"I was playing football in Chamberlain that afternoon, and watched the flight of the balloon throughout the game. I can still remember how we tried to distract the other football team by pointing to the balloon, but it didn't work. They still beat us by 2 or 3 touchdowns. I worked in the barbershop at the Kimball hotel and that evening the lobby was filled with reporters and government officials." [14]

"Margaret Rechtfertig, White Lake's lone telephone operator, was swamped with calls that evening (August 11, 1935) some as far away as New York. Later the National Broadcasting Company purchased the line rights for an hour and a nationwide hookup was originated in the farm home of Mr. and Mrs. Peter Kramer, Sr., which was near the landing site.

William Baumgartner, Western Union operator, was kept busy and a helper was rushed in to aid him. Five thousand pieces of mail from here to every known country in the world were hand-canceled by postmaster Helen Kieffer and her assistant, Alma Bogenhagen. It took them over three hours." [15]

A Mr. Hyde who took Captain Anderson to Kimball Monday night after the landing, learned a few facts about the flight as they drove. Hyde "learned that the balloon inflated was about three times as high as the White Lake Water Tower. At the highest elevation the sky was a deep blue or purple color. To purify the air in the gondola, liquid oxygen was poured on the floor where it evaporated. The inside of the gondola was almost covered with self-registering instruments run by electricity from storage batteries.

The intense cold at high elevations caused moisture to condense on the inside of the gondola, much as moisture gathers on the inside of a windshield in cold weather. The ballast in sacks on the outside were emptied by exploding a dynamite cap placed at the end of each one. The first instrument to be removed (upon landing) and taken east to the U.S. Bureau of Standards was the barograph which gave official figures for the elevation reached. The total weight of the balloon before leaving the ground was 30,000 pounds. There were about 450 pounds of ballast left, which was 50 pounds short of the 500 pounds which Captain Anderson had planned to have at landing. To drop 10,000 pounds in 14 miles with scarcely a jolt was quite a trick in any man's language. Captain Anderson gave his entire attentions to piloting the balloon. To have given attention elsewhere for even ten minutes would have courted extreme danger." [16]

Both pilots, Captain Albert Stevens and Captain Orvil Anderson were given the Distinguished Cross Award for their record-setting probe into the Stratosphere. Captain Albert Stevens retired from the Army as a lieutenant colonel. He died in 1954. Captain Orvil Anderson rose to the rank of major general with a distinguished career in World War II. He died in 1965.

"On Monday afternoon, (50 years later) November 11, 1985, in Bill Ries' pasture in the southwestern part of Aurora County, 12 ½ miles south and 1 ¼ miles west of the White Lake Water Tower, a new monument of native field rocks was dedicated in honor of Captains Stevens and Anderson who made the historic balloon flight of November 11, 1935, and in memory of all those who saw the balloon land on that day but are no longer alive." [17]

The White Lake community conducted its First Annual Balloon Days, August 10, 2010, which honored the 75th anniversary of the historic Stratosphere Balloon landing.

Noting and celebrating historic events such as Explorer II brings honor to the event itself as well as to the people who dare to remember.

Monument was erected and christened with a bottle of water
to commemorate the landing of Explorer II near White Lake.
– Photo courtesy of Candi Walti

Chapter Seven

Stratolab High I & IV

"Above me I saw something I did not believe at first. Well above the haze layer of the earth's atmosphere were additional faint thin bands of blue, sharply etched against the dark sky. They hovered over the earth like a succession of haloes." – David Simons, first balloon flight above 100,000 feet. From "A Journey No Man Has Taken" *Life Magazine*, 2 Sept, 1957.

Once again, in 1956 and 1959, the Stratobowl of South Dakota, would be used to launch 2 balloons in order to penetrate the skies in search of new scientific discoveries. The Stratobowl of South Dakota was a popular launch site because its topographical and natural features provided the protection and cover required for successful inflations and launching of 2 earlier flights–namely–*Explorer I* and *II*. Now *Stratolab I* and *IV* would fly into the stratosphere.

Prior to the launch of *Stratolab I,* during 1951-1953, the Navy developed a new high atmospheric research tool through General Mills and Winzen Research of Minneapolis, called the *Sky Hook Balloon.* The capsules were made by the University of Minnesota and New York University Engineering departments. The *Sky Hook Balloon* project was then used for scientific research flights. During those years, there were 39 aeromedical Field Laboratory Space biology flights in the Midwest, at Holloman Air Force Base and other sites. Live animals such as mice, hamsters, dogs, cats, and fruit flies were tested in the stratosphere. Five of the 39 biology flights where launched from Pierre, South Dakota in late October, 1953.

"After World War II, Soviet dictator Josef Stalin dropped an 'iron curtain' across Eastern Europe, and the U. S. wartime ally became a closed society bent on spreading Communism through-out the world. In an attempt to obtain information about Soviet activities, United States Air Force planners turned to balloons."[1] By mid-1955, plans were made to move from unmanned balloon flights to manned Stratospheric ballooning.

Then October 4, 1957, *Sputnik* shocked the free world! The Soviets launched the world's first artificial satellite. With *Sputnik* the race into space accelerated tenfold.

My neighbor, Custer County resident, Bill Perrett, age 90 then, told a story that made "his hair stand on end" at the time *Sputnik* was circling the globe. I, as a Lutheran pastor, and close neighbor, brought Holy Communion to him from time to time. We talked golf and other subjects. Bill golfed into his nineties. Then unexpectedly, after one of my visits sharing the Lord's Supper, Bill got very excited. He recalled an unforgettable experience with *Sputnik*. He told me that he and his wife were attending the Passion Play in Spearfish South Dakota, which was an outdoor venue chronicling the life and death of Christ. When "Christ" was being lifted up on the cross, a covey of pigeons were released nearby. And just then – a hungry hawk swooped down, attacking the birds, scattering feathers and confusion everywhere, and then, "would you believe," Bill tells, wide-eyed and deeply touched, "*Sputnik* flew overhead to the utter astonishment of everyone in that large gathered audience." For Bill, *Sputnik* flying over at that precise moment, was obviously a frightening and poignant spiritual experience!

The Cold War and *Sputnik* chilled relationships between the Soviet Union and the United States. The development of armaments of all kinds became number one priority for both sides.

I was working on the Bomarc Guided Missile, a defensive deterrent weapon designed to defend against an intercontinental missile attack, for the Boeing Airplane Company in Seattle, Washington. Assigned to plant II, in the engineering department, I worked as a draftsman. This drafting training and experience would later provide an expedient path for my working with balloonist Ed Yost in the late 70's and early 80's.

Whereas in 1935, Explorer II, marked the last high altitude flight and the subsequent completion of the first step of human stratospheric ballooning, it was *Strato-lab I* that ushered in the beginning of the second step of high altitude ballooning and the developing space age, twenty-one years later.

Lee Lewis and Malcolm Ross make final preparations
before *Stratolab I* launches from the Stratobowl.
– Photo courtesy of Jim Winker

It was November 8, 1956, when the *Stratolab I* gondola lifted Lieutenant Commanders Malcolm Ross and M. L. Lewis from the Stratobowl in South Dakota to a new world record for manned balloon flight of 76,000 (23,000 meters). It surpassed *Explorer II's* previous record by 3,000 feet. The purpose of this flight was to provide and utilize a high altitude laboratory, test the physical response of humans within high altitude environments, and conduct specific scientific experiments.

These studies included aero medicine, meteorology, atmospheric physics and astronomy, all of which would provide significant knowledge, regarding the mysteries of outer space.

Patricia Sanmartin, who presently lives next door to the Stratobowl near historic Rockerville, South Dakota, was an eyewitness of both *Stratolab I* and *IV* flights in 1956 and 1959.

Patricia (Koren) Sanmartin, eye witness to both *Stratolab I and Stratolab II* is shown here later as Miss Rodeo USA, 1965.
– Photo courtesy of Patricia Sanmartin

Patricia was twelve years old when the first balloon was launched and recalls going to the rim, seeing the crowds and among them was Senator Francis Case who she remembered because "she had bought a horse from the Case ranch which is across the road from the Crazy Horse

Monument." In a 2012 interview, Patricia remembered the huge, beautiful, silvery balloon ascend and disappear into the Stratosphere. Patricia also recalled *Stratolab IV's* launch. She was there with her boyfriend and was well informed after listening to many stories about *Explorer I* and *II*, witnessed and told by Benjamin Rush. Benjamin Rush was a well known gold miner and timber mill operator at Rockerville. [2]

As the aluminum gondola of *Stratolab I* stretched high in the near vacuum upper stratosphere and the altimeter reached the new record of 76,000 feet, Lee Lewis broke the silence saying, "a lot of very fine people have passed this way." [3]

Stratolab I proved the feasibility of penetrating the Stratosphere with a relatively light, inexpensive man-carrying balloon of polyethylene plastic. The plastic envelopes had been proven earlier in the *Skyhook* flights. "The rubberized-cotton envelope of *Explorer II* – a gigantic 192 feet in diameter when fully expanded – weighted 5,916 pounds with its various accessories whereas comparatively, *Stratolab I* was 128 feet in diameter weighing a mere 595 pounds including valves." [4]

Technician in white outfit and Don Johnson on flatbed in the Stratobowl inflating Stratolab envelope – *Photo courtesy of Jim Winker*

The two aeronauts remained at their peak altitude of 76,000 feet for only a few minutes before starting an unexpected descent, due to a malfunctioning helium valve. Once at a safe level, they started jettisoning equipment out of the portholes to slow their rate of descent. They finally made a safe landing on level, sandy soil near a ranch in Brownlee, Nebraska. Despite the defective valve, the flight was successful in providing new scientific and technical information. "It was a clarion call announcing that the United States was now officially back in business of exploring the stratosphere by means of balloon. President Eisenhower awarded Malcolm Ross and Lee Lewis with the Harmon International Aviation Award in August, 1957, acknowledging their work with the *Stratolab* project." [5]

Malcolm Ross, the physicist and veteran pilot of *Stratolab I*, realized that ballooning was the "logical step toward space astronomy." [6]

On that afternoon of November 28, 1959, when *Stratolab IV* leaped upward from the now famous Stratobowl, little did Patricia, who trotted over from her nearby home and to watch with awe and excitement, realize the significance of it all.

Ross had selected as his co-pilot, Dr. Charlie Moore of General Mills in Minneapolis. Moore served as the scientific observer and was responsible for the inception of the project. "The balloon reached a new record altitude of 81,000 feet. Ross and Moore spent the evening with a 16 inch telescope and spectrograph observing the water vapor in the atmosphere of the planet Venus." [7]

Stratolab IV landed the following afternoon near Manhattan, Kansas. The spherical gondola used in the *Stratolab* flights was retired after this flight.

The three other *Stratolab* flights were not seen by Patricia. They were launched elsewhere, other than in the South Dakota Stratobowl. *Stratolab II* was launched October 18, 1957, from the Feigh Mine at Crosby, Minnesota. The Feigh mine was one of the Portsmouth, Feigh and Managan-Joan mines, complex, in the iron range of Minnesota. *Stratolab II* was a relatively routine and unnoticed flight. Much had changed in the world of aviation and aeronautics since *Stratolab I*. Colonel Joseph Kittinger had set the record (which stood till recently) of the highest parachute jump from the Stratosphere at 102,800 feet. David Simons had broken the 100,000-foot barrier, spending the night in the Stratosphere, and Sputnik had gone into orbit. *Stratolab II* focused on observing instruments to measure the physiological responses of the pilots, and to monitor both mental and physical fatigue.

Stratolab III ventured into the Stratosphere with Ross and Lewis on July 26-27, 1958, from the Portsmouth mine at Crosby, Minnesota.

Stratolab V was launched May 9, 1961 from the deck of the carrier USS Antietam. It was flown by Commander Malcolm Ross and Lieutenant Commander Victor G. Prather. Unfortunately, a tragic climax reminiscent of the fates of earlier aeronauts in history, the 34-year-old Prather fell into the water and drowned while attempting to scramble onto a hook dangling from a rescue helicopter. [8]

Three weeks before the *Stratolab V* flight, the Russian, Yuri Gagarin, on April 12, 1961, became the first man in space. May 5, 1961 Alan Shepherd was launched on America's first suborbital flight – Freedom 7.

The following chart shows the placement and significance of the *Stratolab* flights in context with other Navy and Air Force Stratospheric probes:

Balloon Flight	Date	Pilots	Altitude
Navy *Strato-Lab I*	Nov. 8, 1956	Ross & Lewis	76,000
Air Force *Manhigh I*	June 2, 1957	Kittinger	96,000
Manhigh II	Aug. 19, 1957	Simons	101,500
Strato-Lab II	Oct. 18, 1957	Ross & Lewis	86,000
Strato-Lab III	July 26, 1958	Ross & Lewis	82,000
Manhigh III	Oct. 1959	McClure	100,000
Air Force *Excelsior I*	Nov. 1959	Kittinger	76,400
Strato-Lab IV	Nov. 28, 1959	Ross & Moore	81,000
Excelsior II	Dec. 11, 1959	Kittinger	74,000
Excelsior III	Aug. 16, 1960	Kittinger	102,800
Strato-Lab V	May 4, 1961	Ross & Prather	113,740

With astronauts and cosmonauts orbiting the Earth in the 1960's, the second era of human high-altitude balloon ascents drew to a close. The *Man-High*, the *Excelsior* and *Stratolab* projects helped pave the way for humans to enter space via a new and amazing way.

Chapter Eight

Better than a Circus

"History is the memory of things said and done." − Carl L. Becker

"Joel – do you want to do something better than going to a circus?" I asked my 11 year old son one snowy November day. "What is it Dad?" he said. Joel was a gifted and curious youngster, always looking for an adventure. "It's a gas balloon fiesta in the StratoBowl in the Black Hills." "Let's go Dad, I've got no school on Monday."

I had been invited by Ed Yost to serve as the Chaplain for the 50th anniversary celebration of Explorer II, sponsored by Ed and the National Geographic Society. It was November, 1985.

Next day we packed, left North Mankato, and were on our way. The drive turned out to be a white-knuckled trip going west on I-90. The four-lane was ice packed. My wife had warned us. We were crazy, but we just couldn't miss this unique and historic event whatever it took to get there.

We did make it to the Alex Johnson hotel in Rapid City. Balloonists from all over the country were checking in. Joel and I quickly unloaded, delighting to find ourselves among several "rock star" balloon pilots. One of the National Geographic photographers who was unloading his camera equipment nearby, asked me, "are you one of the balloon pilots?" "No," I answered, "I am the Chaplain for the event." He took a double take, obviously surprised and amused, then turned to his fellow photojournalist and said, "He's the Chaplain."

"Yes, Joel, what we were in for would turn out to be far better than a circus!"

Among those invited were "balloonatics" — *Classical Gas*: owned by Dr. James Jones, a chiropractor, Phoenix, Arizona; *Destiny*: owned by pilot Fred Krieg, Perris, California; *Night Star*: owned by Dewey Reinhard, Colorado Springs, Colorado, private pilot and organizer of the Colorado Springs balloon race; *Ragtime*: flown by Greg Thomson, Moore, Oklahoma, a heavy equipment dealer; *Rodney the Jazz Bird*: (the only one which actually flew), owned by John Shoecraft, Scottsdale, Arizona, who had completed a transcontinental balloon flight; *Benihana*: owned by Ron Clark, Alburquerque, New Mexico, real estate dealer and ballooning partner of restaurateur Rocky Aoka; *Crazy Horse*: owned by Cyril Laan Jr., New Orleans, La., a nail dealer and Dale Yost, Chicago, Ill., advertising executive; *Old Glory*: owned by Robert C. Penny III, Oak Brook, Ill., Penny Family Institute; *City of St. Louis*: Don Caplan, St. Louis, husband of the late Nikki Joyce Caplan, a pioneer woman balloonist and the first person to pilot a balloon through the Gateway Arch in St. Louis, and Jane Buckles, a friend of Nikki Caplan.

Some of the invited **hot air** balloonists were David McPherson, Sturgis, South Dakota; Orv Olivier, Sioux Falls; The Aeronauts, Ltd. (Bernie Tyrell and Russ Pohl), Sioux Falls; Mark Wellenstein, Wynot, Nebraska; Tom Davies, Vail, Colorado; George Albers, West St. Paul, Minnesota; Bob Waligunda, Pittstown, N. J.; Mark Leighton, Hancock, Minnesota; and Wayne Woodmancy, Sharon Springs, Kansas; Ed Yost in the *Pathfinder* as race organizer and director. (Ed Yost's actual title was the Balloonmeister). [1]

Other balloon related notables gathering for this momentus celebration, would be the President of the National Geographic Society, Gil Grosvenor, Ruth Stevens, the 81 year widow of Captain Albert Stevens and daughter Susan, and Tom Mcknew, also of the National Geographic Society.

Ed Yost provided the leadership and following schedule for the two day event:

STRATOBOWL SCHEDULE 1985

1. Two helium trailers are scheduled to arrive Sunday morning, 10 November.
2. Weather and trajectory forecasting will be accomplished by Dr. Emily Frisby.
3. Helium balloon inflation will begin early Sunday afternoon and remain about 90% inflated and reefed down throughout the night.
4. All spectators are invited in the bowl launching area both Sunday afternoon and Monday morning. Crowd exit bowl at 5 p.m. Sunday.
5. No admission charge for spectators either day.
6. 10 November Sunday evening banquet for 200 people hosted by the National Geographic at the Hilton Inn (by invitation only).

5:45 – 6:45

Cash bar

6:55

Guest seated in dining area

Blessing by Rev. Arley K. Fadness

7:00

Dinner Served

7:30

Welcome and Introduction of Dignitaries

Opening address   Gilbert M. Grosvenor,
                                   President National
                                   Geographic Society

Further introduction and
brief historical report   Dr. Tom McKnew,
                  National Geographic Society

Guest Speaker   Malcolm D. Ross
                                   Cdr. USN Ret.

9:00 – 9:30

Conclusion of Program

7. Monday 11 November at launch site:
    0500: balloon inflations to continue
    0615 – 0630: brief ceremony at launch site, such as,

86

a. Military band
b. Color guard
c. Fly-by
d. Any additional addresses required
e. Benediction by Rev. Arley K. Fadness

> 0701: Lift-off of balloons – the same time that Explorer II became airborne.
>
> Sunrise on this date is 0645 MST, with sunset occurring at 1715 CST near-   deathmid-South Dakota. Ten and one-half hours of daylight will allow the balloonists to complete their task.
>
> Ten hot-air balloons will be readied and launched about one hour after the helium balloons depart.
>
> Explorer II landed at 314 pm ten miles south of White Lake, S.D. And Explorer I   crashed on 28 July, 1934 eight miles northwest of Holdrege, NE. The balloon that lands closest to either of these two targets will be declared winner of the event.

The program was to feature Commander Malcolm D. Ross, 65, retired USN, as the banquet speaker. Ross was the perfect choice. He was an atmospheric scientist and premier balloonist. As a participant in the *Stratolab* series he flew out of the Stratobowl, November 28, 29, in 1956 and in 1959 in *Stratolab IV*. He set one stratospheric ballooning record, when on May 4, 1961 he flew the highest manned balloon flight reaching 113,740 feet. Unfortunately his partner, Lt. Commander Victor A. Prather, drowned while being rescued at sea at touchdown.

Then a shocker, Malcolm Ross died October 8, 1985, a month and two days before his scheduled banquet talk in Rapid City. Other speakers filled in and made the banquet that Sunday evening, November 10, a most memorable and magical evening.

As the Chaplain, I offered the following evening banquet prayer:

> *"O Lord, our Lord, how Majestic is your name in all the earth.*
> (Psalm 8:1–RSV)
> *By your hand you fashioned the earth, the seas and the heavens.*

Help us this night to celebrate the mysteries and glory of Creation. We see beauty, order and harmony and with the eyes of faith recognize your Divine imprint.

We are thankful that you created man and woman in your image—- capable of so many things—of scientific inquiry, of daring sport, able to sing and dance, and to love and to pray.

Tonight, we especially celebrate people who chose to be more than hewers of wood and drawers of water — people with inquiring minds and courage in their spirits to probe the high adventures of the stratosphere.

Bless us who inhabit the troposphere, who believe we are here by choice or by chance.
Enable us to be diligent in our work, excited about our dreams, and faithful in that which gives us meaning.

Bless those whose generosity made this event possible.
Bless our food this night, and the farmers who grew it.
Bless our conversations and good friendships through Christ our Lord. Amen."

Next morning, the events at the Stratobowl began to both develop and unravel as the snow and the wind threatened to deny the best laid plans. A journalist once described the bowl as though "someone used a giant cookie cutter to lift out a huge chunk of earth leaving a football field-sized amphitheater surrounded on three sides by almost vertical cliffs." [2]

The bowl became a winter wonderland, beautifully clothed in white, serene, protected from the nasty upper winds. It belied the danger beyond the sheer walls surrounding it. A carnival atmosphere warmed the spectators as the 7th Cavalry Band began to play. Joel and I imagined the Stratobowl a Circus Big Top. The balloon pilots and crew were the acrobats, and clowns, while Ed Yost cracked the whip, as the Ring Master announcing "The Greatest Show on Earth!"

Nine gas balloon envelopes laid out on the bowl floor like nine buttermilk pancakes. They were jam red, blackberry black, banana cream beige, maple syrup brown and many assorted tasty colors. Two were inflated–Fred Krieg's *Destiny* and John Shoecraft's strawberry red, *Rodney the Jazz Bird*. "Right now there's a 50-50 chance for launch," Yost said. "Eagerness wants to go, common sense says stay here." The plan was to inflate all the balloons with helium at a cost of $2,000 per balloon.

It became clear that the high winds would force a delay and probable postponement of any commemorative flight. Yet the pilots and Ed hoped for a subsiding of the strong south winds by mid-morning. John Shoecraft's *Rodney the Jazzbird* made several 30 foot tethered ascents in symbolic takeoffs to commemorate the Explorer II flight. Both *Rodney* and *Destiny* gave brief rides to several of the attendees. Gil Grosvenor, and son Graham, Ruth Stevens, and others, enjoyed the tethered rides. "It was wonderful," said Mrs. Stevens, who was at that time in the process, researching and planning to write her book, *My Husband, the First Astronaut*.

Finally it became clear that the launch needed to be postponed. Ed remarked, "Once you clear that ridge and get out of here, you'd be moving at 50 miles per hour, and who is going to land at 50 miles per hour? Who would want to jump out of a moving car at 50 miles per hour?"

The hot air balloon flights were also canceled. Soon the pilots began to pack their balloons. At noon Fred Kreig deflated his big blue and yellow balloon *Destiny*.

Only John Shoecraft's balloon remained inflated as he hoped for a lessening of the winds. Then shortly after 1 p.m. the winds aloft were down to 20 mph. John Shoecraft decided to launch and make the flight. Tension mounted among his ground crew as they busily prepared the red bird for flight.

Ed gave me the high sign – it was time for the launch benediction. I said, paraphrasing Psalm 100:

> *"Make a Joyful Noise unto God all you lands;*
> *Sing forth the honors of God's name;*
> *Make his Praise glorious.*
> *Bless all creatures of the land and of the air–*

feathered eagles
gorgeous butterflies
angels and seraphims
and yes, aviators and aeronauts.
On this special day in honor of history and science,
bless, keep, and protect men and women of courage and adventure.
We give thanks for this great Land and all lands where justice
and righteousness prevails.
We pray for good weather, a speedy launch and safe landing.
Grant to all veterans (on this veteran's day) our gift of thanks
and give to us all that peace that passes all understanding both
in our world and in our hearts,
through Christ who is our Prince of Peace.
Amen."

Chaplain Fadness with son Joel, prior to
the launch of Rodney the Jazz Bird

Ed Yost shook hands with John Shoecraft, smiled, and said, "I think
you are going to win the race. Have a good ride." Up, up and away,

flew *Rodney the Jazz Bird*. The flight would be a short but successful endeavor — landing safely near the Rapid City airport, some 15-20 miles from launch.

As the balloon soared over the 450 foot eastern rim, Ed said, "It's a heck of a deal. The wind has really changed. We could have gotten them all off had we known this." [3]

The 50th anniversary celebration of Explorer II may not have been as successful as planned, yet it did bring together swarms of spectators, premier gas balloon pilots and a symbolic win for all. Eleven year old son Joel, agreed, "it was better than a circus!"

The National Geographic Society's plan for journalistic coverage of the 50th anniversary commemoration of Explorer II's flight out of the Black Hills Stratobowl, didn't fully materialize in the Society's March magazine edition. [4]

Nonetheless, President Gilbert Grosvenor of the National Geographic Society, wrote in the following March, 1985, issue, "It became a matter of pride to Ed and to all of us to try to launch at least one symbolic balloon in the face of howling winds, blowing snow and icy mountain roads. The big helium trucks worked down the drifted switchbacks, and somehow John Shoecraft inflated his big red balloon, *Rodney the Jazzbird*. While Captain Shoecraft waited on the weather, he gave tethered short flights to Ruth Stevens, widow of Captain Stevens, then to me and my son Graham–in hopes Graham will remember it in 2035.

When it cleared briefly, John launched. The sight of that lone red sphere rising against the snowy pines and rocky cliffs reminded me of the business we are in—pushing back the frontiers of knowledge often in the face of adversity." [5]

The first celebration of Explorer II's flight took place on it's **10th anniversary,** November 10, 1945. A banquet was held at the Alex Johnson hotel with the following notables attending: Dr. Gilbert Grosvenor, President of the National Geographic Society, Dr. Thomas W. Mcknew, Secretary of the National Geographic Society, Dr. Lyman J. Briggs, Director of the National Geographic Society, Major General Curtis LeMay, Commanding General of the 20th Air Force, Melvin M. Payne, Assistant Secretary of the National Geographic Society, Frederick

Simpich, Assistant Editor of the National Geographic Society, and Richard Stewart, Staff Photographer, National Geographic Society.

On Sunday morning, November 11, a worldwide radio hook up enabled speakers such as Lt. Colonel Stevens from San Francisco, Major General Anderson from Tokyo, Major General Kepner from Berlin and others to speak. As part of the ceremonies a facsimile of a bronze plaque, which was in the process of being made, was placed upon a boulder on the rim of the Stratobowl. [6]

Joe Kittinger, (third from left) the featured speaker at the 25th Explorer II anniversary. (from left to right) Maude Hill, Queena Stewart, Joe Kittinger, Belle Stewart and Bob Hill – *Photo courtesy Helen Wrede*

On the **25th anniversary** of the Explorer II flight, the National Geographic Society, in conjunction with United States Air Force, decided to honor this event on November 10 and 11 of 1960.

Captain Joe Kittinger, just back from his historic parachute jump from a balloon at 102,800 feet, was one of the primary speakers at the bowl. Later he remembered the evening of November 10th at the Alex Johnson hotel in Rapid City when he said, "The evening of the 10th of November was spent in Rapid City at the Hotel where we attended a grand banquet celebrating the Explorer flights. The dinner was attended by all of the dignitaries, including the Mayor of Rapid City. There were many people at the dinner that had attended the Explorer I and Explorer II flights. I'll never forget one man's descriptions of the balloon failure following the Explorer I flight. He said that he was up on the rim of the

Stratobowl watching the balloon being inflated. It was a cold evening and to fortify himself, he was sipping on a bottle of moonshine as he watched the inflation of the balloon over the outline of the liquor bottle. Suddenly, the balloon ripped and fell to the ground. He thought that the bottle had broken. It was quite funny the way that he described the incident.

General Kepner also described how he selected the Stratobowl as the site for the balloon flights. He was the Army Project Officer for the Explorer Flights and was notified of a deep Canyon near Rapid City. He traveled there on horseback with his cowboy guide and rode to the rim of the Canyon saying, 'God made this site from which to launch balloons.' It was a grand evening of nostalgia and funny stories about the goings on at the various Explorer flights." 7

The next day, November 11th, Joe Kittinger,, was one of the prime speakers at the Stratobowl rim. Ed Yost was down below in the bowl preparing for his 2nd historic hot air balloon flight. Seated at the rim were Dr. Lyman Briggs, chairman of an Advisory Council for the 1935 flight and Director Emeritus, National Bureau of Standards; Lt. General Homer Jensen, State Adjutant General; Lt. Gen William E. Kepner, one of the three pilots of Explorer I; Dr. Thomas Mcknew, Secretary and Executive Vice President of the National Geographic Society; Lt. General Albert T. Wilson, Deputy Comptroller, U.S. Air Force; Lt. Colonel David G. Simons, who ascended in a sealed capsule to more than 101,000 feet; and General Orvil Anderson, one of two pilots of Explorer II. Albert Stevens the other pilot had died in 1949. 8

Overhead, a U-2 aircraft called a message to the assembled group on the rim by VHF radio.

Speeches were given by Dr. Mcknew, Generals Kepner and Anderson. Joe Kittinger remembers, "The ceremony lasted one hour and was quite an historic event. Little did we know that in a few years, man would fly in space and then after another few years man would walk on the moon. But it all started right here at the Stratobowl." 9

Five years later, in 1965, the **30th anniversary** celebration featured a return visit by General Kepner and General Anderson, both then retired. Also present were Dr. Thomas Mcknew, Dr. Melvin Payne, Senator Karl Mundt, Senator Kerr and several National Geographic Society officers and trustees. At a reception held in honor of the special guests, Dr. Hugh Dryden, Director of the National Advisory Committee for Aeronautics, told of the research that had developed as a result of the flights of Explorer I and Explorer II. Captain Joe Kittinger was back, and told of his parachute jump from 102,800 feet. He told of the training, research, emotions, thoughts and ultimate satisfaction he had experienced. 10

October and November, 2005, was the time to celebrate the **70th anniversary** of Explorer II.

The Journey Museum in Rapid City, Ray Summers, Executor Director, and Reid Riner, Curator of the Minilusa Historical Association, presented a program from October 2 to November 13, entitled, *"The Black Hills Journeys Into the Space Age."* The Stanford Adelstein Gallery displayed an exhibit of Stratobowl artifacts. Programing during this time featured a "Storyteller Series." The Series included the following storytellers, stories and one field trip:

October 9 – Kevin Kuehn, Rapid City balloonist, *"I flew from the Stratobowl."*

October 16 – Charles Gannon, *"The Legacy of Albert Stevens, Aviation and Photography Pioneer."*

October 23 – James Winker, Chief Officer, Raven Industries, *"The Influence of Stratobowl Technology on Modern Ballooning."*

October 30 – Panel of "Eyewitnesses Discussion" of the launches of Explorer I and II.

November 5 – Bus tour to the Stratobowl – Helen Wrede and Arley Fadness, hosts.

November 6 – Arley Fadness, Historical Balloon Society, *"Up, Up and Away: Celebrating the Stratobowl."*

Celebrating the 70th anniversary of Explorer II at the Journey Museum.
– *Courtesy of the Journey Museum*

94

The Stratobowl Exhibit Planning Committee under the direction of the Minnilusa Historical Association, included Wini Michael, Judy Cobb, Karen Miller, Arley Fadness, Reid Riner, Helen Wrede and other history enthusiasts.

In September, 2010, the **75th Diamond Jubilee anniversary** celebration of Explorer and the 50th anniversary of hot air ballooning, took place at the Stratobowl. Mark and Kay West and Ken and Cory Tomovich met with the Stratobowl historical Committee, and planned various ballooning events. Mark West was the former President of Raven Industries, Aerostar, and new Chief Technology Officer for Raven. Kay West, was the owner and operator of *Prairie Sky,* Inc. and Event Organizer. The Tomovichs are one of the private land owners of the bowl. The first annual Stratobowl celebration hot air launches, took place September 24-26, consisting of hot air balloon flights out of the Stratobowl, balloon glows at sunset, and a picnic.

Hot air balloon ascension out of the Stratobowl marking themselves 75th anniversary
– Photo courtesy of R. Shipporeit

The second annual Stratobowl celebration was held September 9-11, 2011. "Several hot air balloons launched from the Stratobowl at sunrise each day. There was also a balloon glow Saturday evening just after dark when the moon began to rise." 11

P. S. A third celebration took place in 2012 and the tradition will continue in 2013.

Joel agreed. "A balloon celebration/ascension is <u>better</u> than a circus." But why not go to <u>both?</u>

Chapter Nine

A Lasting Legacy

"A country with no regard for its past will do little worth remembering in the future." – Abraham Lincoln

When daring men and women dream audacious dreams and the dreams blossom, they leave a legacy to be remembered and celebrated forever.

Consider Doane Robinson, South Dakota State Historian. Doane Robinson, while reading of Gutzon Borglum's sculpture on Stone Mountain, Georgia, was inspired to launch an ambitious project. He dreamed of sculpting the granite pillars of the Needles of the Black Hills into likenesses of famous people. Why not create an attraction that would magnetically draw tourists from all over the globe to South Dakota? Doane Robinson, enlisted the interest and help from Gutzon Borglum. After considerable struggle, they secured funding along with support from key allies as John Boland, Congressman William Williamson, Senator Peter Norbeck and President Calvin Coolidge. They switched the location of a carving from the Needles to what is known as Mount Rushmore near Keystone. Four presidents would be blasted into granite, forever memorializing Washington, Jefferson, Roosevelt and Lincoln. Doane Robinson's dream would see as many as that record of 17,600 visitors on August 13, 2010.

Another dreamer was Korzak Ziolkowski, who at the request of Chief Henry Standing Bear and other Chiefs of the Lakota Nation, agreed to carve the largest monument in the world, near Custer, South Dakota. The purpose was to honor Chief Crazy Horse and all Native Americans. On June 3, 1948, the first blast signaled the beginning of an amazing project that continues to the present day. "When will the monument be finished?" many ask. Ruth Ziolkowski, the wife of Korzak and the present CEO says, "we just keep making progress." One of the popular events surrounding the blasting and carving of the mountain is the annual volksmarch. In June of 2012, over 11,000 volksmarch hikers climbed to the top of the arm of the carving for the annual trek set on the first weekend in June. Daily attendance and contributions grow as special events and carving progress continues on the mountain, blast by blast.

E. Paul Yost was another dreamer whose vision grew out of his own personal experimental adventures and accomplishments. (See Raven Rocks – chapter 13, Crossing the Pond – chapter 16). Ed, called the "Father of the Modern Hot Air Balloon," dreamed of a National Monument that would celebrate the beginning of the Space Age at the Stratobowl near Rapid City. Historic stratosphere flights in *Explorer I* and *II* and later, 2 *Stratolab* flights, deserved public notice and celebration, Ed believed. Ed said, "I hope for an improved road, similar to the Iron Mountain Road that would wind through the trees....and we should have a building with everything in it." The everything would include information and items from the *Explorer I* hydrogen-filled balloon that drifted up out of the Stratobowl early on June 28, 1934, as well as the *Explorer II* helium-filled balloon that followed in 1935." [1]

The ultimate artifact would be the return of the actual *Explorer II* gondola which is presently housed at the Smithsonian Institute in Washington, D. C.

Was Ed Yost's dream of an interpretative center/museum positioned near highway 16 and staffed during the tourist season unrealistic? Bob Hayes, local historian and columnist, in preparation for a Dakota Conference presentation, sent out a general inquiry asking the question, "Why should we make an interpretative park at the rim of the Stratobowl?" Bob then made the case quoting several key proponents. "Colonel Joe W. Kittinger, ASAF (Retired) responded by saying, 'My answer is this: The Stratobowl flights in 1934 and 1935 were the beginning of our Space Program. We owe it to the citizens of our country to preserve this history and make it available to anyone who would like to visit the rim of this historic site. Thirty-seven years before the first Mercury flight blasted off at Cape Canaveral, balloons lifted majestically from the floor of the Stratobowl. Our nation needs to honor this location for the beginning of space exploration, as our nation has been the world's leader in this quest to explore space. This is just one more reminder what a great nation America is, and the greatness of the early pioneers either on the ground, in the air, or in space.'" [2]

High interest and welcome support came from the local citizenry such as Jim Shaw, former mayor of Rapid City, John Twiss, former Supervisor of the National Forest Service, and many living eyewitnesses of the Explorer I and II flights.

Jessie Y. Sundstrom, well known Custer historian and author of several books and articles, as well as former publisher of the Custer Chronicle for many years, responded to Bob Hay's question on March 13, 2007 as follows:

"Regarding the Stratobowl, my own mother was present for those flights. She slept out on the rim during one of those flights. I am sure she would approve of the idea of having a park on the rim. We always felt kind of cheated when it was blocked off to the public. The Stratobowl area should be managed as a park to honor the pilots and scientists of an endeavor that brought this country to where it is at present in the space program. In keeping with the importance and dignity of the project and the men who carried it out, the park should have a patriotic atmosphere befitting the success, but also the world of adventure. It should be a federal project since it has benefited, not only the United States of America, but also the world. The park should be policed in some way." [3]

In an interview and working with Helen Wrede, 92, of Rapid City in 2011, I learned of her keen involvement and interest in Ed's dream. Her father, Web Hill, was the president of the Rapid City Chamber of Commerce and worked personally with the Explorer crews. Helen's family hosted and entertained the pilots, Kepner, Stevens and Anderson, in their home. Helen says, "Before the history of this important event has been forgotten completely, a park should be developed and staffed, particularly in the summer months, to let people know the significance of what took place here. In the early 1920's when ideas were beginning to form about carving Mount Rushmore, Doane Robinson asked the question, 'Would it be possible?' A park on the rim of the Stratobowl certainly wouldn't be the magnitude of Mount Rushmore, but would memorialize the beginning of the Space Age."

Reid Riner, historian and Curator of the Minnilusa Pioneer Museum, in conjunction with the Rapid City Journey Museum, has worked and directed several Explorer I and II commemorations and agrees that Ed Yost's dream in some manner should one day become a reality.

Several groups and individuals were less than enchanted with Ed's dream. "Not everyone believes developing a paved road, parking lot and museum on the rim above the Stratobowl on National Forest Service land would be a good idea." [4]

The National Forest Service personnel resisted and said "no" to any development of the land into any kind of interpretative center on public land held in trust for the common good. According to their mission, their role is not to develop memorials on Forest Land.

A letter to me (Fadness) from John Cooper of the State Game and Fish and Parks Department, dated August 1, 2006, stated "Your interest in developing an improved road, parking facilities and interpretive center is not consistent with the Decision Notice and our department's management plan for bighorn sheep in the area. Such development would disturb an important sheep lambing area and would be detrimental to the future of the bighorn sheep herd that the Black Hills National Forest Service and South Dakota Department of Game, Fish and Parks have worked cooperatively on making a success for the last 20 years." [5]

"Yost's big dreams would be bad news for the Stratobowl and people living nearby" said one resident in the bowl. "If the idea would succeed one would likely bring legal action if it looked as if Yost's plan might succeed." [6]

Frank Carroll, spokesman for the Black Hills National Forest, said, "We're not in active negotiation or even talking about this as an agency. He's (Yost) talking about quite an enterprise. But I grew up in a time of giants. And Ed Yost is a giant in many ways, one of those giants with wild ideas and dreams. And I never try to marginalize somebody's dreams and ideas." [7]

Receiving great practical support for programs and learning events as the moon walks, from Amy Ballard and others, Ed's dream, none the less, caused justifiable consternation for the National Forest Service personnel.

In a personal letter to me (Fadness), Ed Yost said, "Don't expect miracles from our combatants-success might require many moons."

Next, Ed contacted private land owners in order to determine their willingness to sell 7 acres of zoned park forest land next to the highway. The Taylors owned an ideal setting, however, I was informed, as was Ed, that the forest on that land was designated as a Tree Conservancy and the Taylors intended to leave the land to their family inheritors.

A letter dated May 25, 2004, from the governor's office in Pierre, politely indicated no real interest in Ed's dream. Governor Rounds gave minimal attention to my letter, passing off the issue to the Office of Tourism. Several personal trips I made to the Office of Tourism resulted in a pleasant drive to Pierre and nothing more.

In 2002, The Balloon Historical Society was founded "to bring public attention and historical recognition to U. S. Stratospheric balloon flights that investigated the cosmos and led the way to space exploration and travel." The early directors of the Society were Paul E. (Ed) Yost, Founder and President of Vadito, New Mexico; Christine Kalakuka, Secretary, from Oakland, California; Bill Gibson, Legal Counsel of Vadito, New Mexico; and Arley K. Fadness, Liaison of Activities from Custer, South Dakota. Later, Brent Stockwell from Oakland, California, would replace Christine as Secretary and Christine would become the Historian. John Craparo, BHS member, was also a special supporter. The first project was to memorialize the very important scientific flights that took place from the Stratobowl in 1934, 1935, 1956, and 1959.

This small commemorative plaque attached to a large boulder on Forest Service Land at rim of the Stratobowl placed in 1955 by the National Geographic Society.

Phase One of the Stratobowl recognition, consisted of designing, fabricating and installing three informative granite historical monuments at the north rim, controlled by the Black Hills National Forest, Mystic Ranger District, Bob Thompson, Ranger. Permission, after some haggling, was granted. The stones six feet tall and four feet wide were constructed by the Bell Monument Company of Beloit, Kansas, and installed. The stones' inscriptions furnished the historic and scientific facts regarding the events in the Stratobowl. Funding was supplied by Ed Yost at a cost of between $60,000 to $100,000.

(From Left to right) Nicole Yost, Ed Yost, Joe Kittinger and photographer watch Jim Bell and worker place the first stone marker.

Later, a fourth stone was added, even though the Balloon Historical Society had promised Ranger Bob Thompson that there would be no further development.

The concrete bases were built by Tony Jenniges of Custer, and Ed Yost. Those present at the installation of the granite markers on one early October day, 2003, were Ed Yost, Nicole Yost, Ed's granddaughter, Joe and Sherry Kittinger, myself and Jim Bell of Bell Memorials.

Ed Yost and first three markers

The stone monuments were dedicated on the **70th anniversary** of Explorer I on July 28, 2004, at the Stratobowl rim. They tell the story of the Stratobowl flights. A launch of the Smokey Bear Balloon on the Jorge and Patricia Sanmartin nearby land, by pilots Bill Chapel and Jim Russell from New Mexico, was canceled, due to unfavorable weather. However, the dedication proceeded with music by the 7th Cavalry Drum and Bugle Corps, dedication and comments by Fadness, speeches by Rapid City Mayor Jim Shaw, Ed Yost, Frank Carroll for John Twiss, and the celebrative speaker – Col. Joe Kittinger Jr. My morning Invocation set the tone:

"O Lord, our Lord, how Majestic is your name in all the earth. By your hand you fashioned the earth, the seas and the heavens. This morning we celebrate people who chose to be more than hewers of wood and drawers of water—-people with inquiring minds and courage in their spirits, to probe the high adventures of the stratosphere and space. We recall names like Kepner, Anderson and Stevens, names like Chuck Yaeger, Jackie

Robinson, Neil Armstrong, Martin Luther King, Sally Ride and Kittinger and Yost.

Bless us who inhabit the troposphere, who believe we are here by choice or by chance.

Enable us to be diligent in our work, excited about our dreams, extravagant in love and faithful in that which gives meaning.

Bless us all as we dedicate these markers so children may learn, historians contribute and researchers be informed.

In your name we offer these petitions, through Christ our Lord. Amen."

The four markers and Old Glory

The texts, of the four granite monuments, read as follows:

<u>**Monument One:**</u>   **Introduction**

OUR CIVILIZATION OCCUPIES A LARGE
SPHERICAL PLANET WHICH HAS A
DIAMETER OF 8,000 MILES—THEREFORE A
CIRCUMFERENCE OF 24,000 MILES.
IT ROTATES ONCE EVERY 24 HOURS, AND HAS
A VELOCITY OF 1,000 MILES PER
HOUR AT THE EQUATOR.

SURROUNDING AND REVOLVING WITH THE
EARTH IS A LAYER OF AIR (78%
NITROGEN, 21% OXYGEN, TRACES OF
MOISTURE). HUMANS CAN FUNCTION
NORMALLY BELOW THE 12,000 FOOT LEVEL,
BUT NEED ADDITIONAL BREATHING
OXYGEN AT HIGHER ALTITUDES.

AT ALTITUDES ABOVE 38,000 FEET, FLUIDS IN
THE BODY WILL VAPORIZE UNLESS
THE BODY IS KEPT UNDER PRESSURE.

HIGH ENERGY PARTICLES FROM SPACE
CONTINUALLY BOMBARD THE
ATMOSPHERIC SHIELD ABOVE US.

IN ORDER TO EXAMINE AND STUDY THESE
PHENOMENA, A LARGE BALLOON WAS
CONSTRUCTED. IT CARRIED A METAL SPHERE,
PRESSURIZED WITH A BREATHABLE
AIR MIXTURE. THIS GONDOLA WAS A
LABORATORY FOR SCIENTISTS.

THE REGION ABOUT 13 MILES ABOVE THE
EARTH WAS SELECTED FOR THE FIRST

EXPLORER ASCENT. ITS PARTIAL SUCCESS
PROVIDED KNOWLEDGE THAT MADE
THE EXPLORER II FLIGHT FLAWLESS IN 1935.

DURING THE 4 ½ HOUR CLIMB TO ALTITUDE,
THE THINNING AIR ALLOWED THE
BALLOON ENVELOPE TO CHANGE FROM A
SLENDER, ELONGATED SHAPED TO A
LARGE, ROUND BALLL. AT MAXIMUM
ALTITUDE THE BALLOON EXPANDED TO 192
FEET IN DIAMETER. (32 FEET GREATER THAN
THE WIDTH OF A FOOTBALL FIELD!)

A LARGE AMOUNT OF LEAD SHOT BALLAST
WAS CARRIED:
TO INCREASE THE RATE OF CLIMB;
TO SLOW THE DESCENT RATE;
TO INDUCE LEVEL FLIGHT.

WEIGHT SCHEDULE

BALLOON & RIGGING:	*6,500*
GONDOLA WITH PAYLOAD:	*4,500*
BALLAST	*4,000*
GROSS WEIGHT	*15,000*

(7 ½ TONS)

DIAMETER OF GONDOLA: 9 FEET

Monument Two:

THOUSANDS OF FASCINATED SPECTATORS
SURROUNDED THESE
LIMESTONE CLIFFS ALL NIGHT TO WATCH
LAYOUT, RIGGING, AND
INFLATION OF EXPLORER I BY A SMALL ARMY
OF WELL-TRAINED AND
DEDICATED INDIVIDUALS. AT SUNRISE, 28
JULY 1934, THE GIGANTIC
BALLOON CLIMBED SILENTLY
FROM THE BOWL.

THE PROJECT WAS ENVISIONED AND
ORGANIZED BY GILBERT
GROSVENOR, PRESIDENT OF THE NATIONAL
GEOGRAPHIC SOCIETY,
IN CONJUNCTION WITH THE ARMY AIR CORPS,
FORT MEADE CAVALRY
TROOPERS AND THE SOUTH DAKOTA
NATIONAL GUARD INFLATED THE
BALLOON. DOW CHEMICAL COMPANY,
BUILDER OF THE GONDOLA,
AND INDIVIDUALS, CORPORATIONS, AND
LABORATORIES INTERESTED
IN ADVANCING KNOWLEDGE PARTICIPATED.

THIS ENORMOUS LABORATORY WEIGHED SIX
TONS. THE 3,000,000
CUBIC FOOT ENVELOPE, FABRICATED BY THE
GOODYEAR ZEPPELIN
COMPANY FROM ALMOST 3 ACRES OF
RUBBER-COATED COTTON,
WEIGHED 5,000 POUNDS. THE BALANCE OF
WEIGHT WAS THE
GONDOLA, CREW, BALLAST, BATTERIES,
SCIENTIFIC EQUIPMENT,

AND LIFE-SUPPORT SYSTEMS.

AS THE LARGEST BALLOON EVER BUILT
CLIMBED THROUGH 15,000
FEET, MAJOR WILLIAM KEPNER ORDERED
THE HATCHES CLOSED.
CAPTAIN ALBERT STEVENS AND LIEUTENANT
ORVIL ANDERSON
ACTIVATED LIFE SUPPORT SYSTEMS AND
DEPLOYED AND MONITORED
THE FANTASTIC ARRAY OF INFORMATION-
GATHERING EQUIPMENT.
AT 40,000 FEET THE GEIGER COSMIC RAY
COUNTER WAS TURNED ON
AND BEGAN CLICKING.

*AT NOON MORE BALLAST WAS DROPPED
TO INCREASE THE RATE OF CLIMB. ONE
HOUR LATER, 11 MILES ABOVE THE EARTH,
A NOISE WAS HEARD AND A LARGE RIP
APPEARED IN THE BOTTOM OF THE
ENVELOPE. THE BALLOON STARTED
DOWNWARD, GATHERING MOMENTUM. IT
FELL 700 FEET PER MINUTE AT 20,00 FEET,
THEN DISINTEGRATED AT 8,000 FEET,.
ANDERSON PARACHUTED TO 5,000 FEET,
STEVENS AT 3,000 FEET, AND KEPNER JUST
500 FEET ABOVE THE GROUND. NO ONE
WAS INJURED.*

*THE SPHERICAL GONDOLA CREATED A
SMALL CRATER IN A SOUTH CENTRAL
NEBRASKA CORNFIELD, 350 MILES
FROM WHERE THE FLIGHT BEGAN TEN
HOURS EARLIER.*

Monument Three

THE MALFUNCTION OF EXPLORER I WAS
CAUSED BY IMPROPER
PACKING OF THE ENVELOPE SO IT DID
NOT UNFOLD
COMPLETELY. AS A RESULT, THE APPENDIX
DID NOT DEPLOY,
THERE WAS NO AVENUE OF RELEASE FOR
THE HYDROGEN
GAS AS IT EXPLANDED, AND THE ENVELOPE
RUPTURED. THIS
PROBLEM WAS SOLVED IN THE
NEXT ENVELOPE.

EXPLORER II

CHANGES IN DESIGN AND OPERATION FOR
EXPLORER II
INCLUDED THE USE OF HELIUM RATHER
THAN FLAMMABLE
HYDROGEN. SINCE HELIUM IS SLIGHTLY
HEAVIER, THE
VOLUME OF THE ENVELOPE WAS INCREASED
TO 3,700,000
CUBIC FEET TO ATTAIN THE DESIRED
FLOATING ALTITUDE.

*THE NEW SPHERICAL GONDOLA WAS INCREASED
TO NINE FEET IN DIAMETER TO ACCOMODATE
SCIENTIFIC EQUIPMENT.*

*AT 7:01 A.M. ON NOVEMBER 11, 1935, TWO
UNITED STATES ARMY AIR CORPS OFFICERS -
CAPTAIN ALBERT STEVENS AND CAPTAIN ORVIL
ANDERSON –LIFTED OFF IN THE GIANT
EXPLORER II BALLOON AND SET OUT ON A*

FLIGHT THAT WOULD TAKE THEM INTO THE STRATOSPHERE TO A HEIGHT NO HUMAN BEING HAD YET REACHED —- 72,395 FEET (13.7 MILES).

THE TRAJECTORY WAS DUE EAST AND THE BALLOON STAYED AIRBORNE FOR 8 HOURS AND 13 MINUTES, TRAVELING 255 MILES BEFORE MAKING AN EGGSHELL LANDING 12 MILES SOUTH OF WHITE LAKE, SOUTH DAKOTA. A MARKER WAS ERECTED NEAR THE LANDING SITE ON THE NORTH SIDE OF THE ROAD.

SOME OF THE EXPERIMENTS CONTAINED IN
THE TON OF
SCIENTIFIC EQUIPMENT ON BOARD
EXAMINED COSMIC RAY
ENERGY, OZONE DISTRIBUTION, SPECTRA/
BRIGHTNESS OF
THE SUN AND SKY, CHEMICAL COMPOSITION,
ELECTRICAL
CONDUCTIVITY AND LIVING SPORE CONTENT
OF THE AIR.

FLIGHT HELMETS DID NOT EXIST SO THE
CREW BORROWED
TWO FROM THE RAPID CITY HIGH SCHOOL
FOOTBALL TEAM
AND THOSE HELMETS EXIST TODAY.

Monument Four

LIGHTWEIGHT PLASTIC FLIM, DEVELOPED IN
THE FIFTIES WITH
FUNDING FROM THE U. S. NAVY OFFICE OF
NAVAL RESEARCH, ALLOWED
FOR LARGER, LIGHTER BALLOON
ENVELOPES. THOSE FOR ROUTINE
UNMANNED FLIGHTS GAINED IN SIZE, AND
CARRIED SCIENTIFIC
PAYLOADS TO MUCH HIGHER FLOATING
ALTITUDES. BETWEEN 1950
AND 1960 TREMENDOUS
ADVANCEMENTS OCCURRED,
ALLOWING FOR THE FOLLOWING FLIGHTS.

IN SPEPTEMBER 1959 DR. MARCEL SCHIEN OF
THE UNIVERSITY
OF CHICAGO USED A 6,000,000 CUBIT FOOT
BALLOON AT
A FLOAT LEVEL OF 153,000 FEET WITH
A 150 POUND
SCIENTIFIC PAYLOAD TO INVESTIGATE
ANTIMATTER.

IN NOVEMBER 1961, DR. MASATOSHI
KOSHIBA OF THE
UNIVERSITY OF TOKYO UTILIZED A 10,000,000
CUBIC FOOT
BALLOON IN SOUTHERN CALIFORNIA THAT
FLOATED AT
112,000 FEET AND LANDED 40 HOURS
LATER IN NORTH
CAROLINA. THE 2200 POUND SCIENTIFIC
PAYLOAD RECORDED
NATURE'S ATOM SMASHING.

STRATOLAB

THE NAVY'S MANNED STRATOLAB
FLIGHTS CARRIED A
SPHERICAL PRESSURIZED GONDOLA. ON 8
NOVEMBER 1956,
WITH A POLYETHYLENE BALLOON OF 800,000
CUBIC FEET,
COMMANDERS MALCOLM ROSS AND LEE
LEWIS LAUNCHED FROM
THE STRATOBOWL, FLOATED AT 76,000 FEET,
WHILE COLLECTING
RADIATION DATA. DEPARTING FROM THE
STRATOBOWL
ON 28 NOVEMBER, 1959, COMMANDER ROSS
AND VETERAN
BALLOONIST/SCIENTIST CHARLIE MOORE
HAD A TELESCOPE
MOUNTED ON TOP OF THE
GONDOLA FOR ASTROPHYSICAL
RESEARCH. THEIR
2,000,000 CUBIT FOOT BALLOON FLOATED AT
82,000 FT.

HOT-AIR BALLOONS

12 NOVEMBER 1960, A U. S. NAVY
EXPERIMENTAL HOT AIR
BALLOON, FLOWN BY
CREATOR AND BUILDER PAUL
"ED" YOST MADE
ITS SECOND FREE ASCENSION. THE FLIGHT
ATTAINED 9,000 FEET AND LASTED
TWO HOURS.

The Balloon Historical Society, having accomplished a significant step towards "The Dream," disbanded upon the unexpected deaths of Christine Kalukaka and later in 2007, the sudden death of Ed Yost. Will there ever be an interpretive/educational center or on-going special recognition of the historic, scientific events that took place at the Stratobowl?

In 2011, John Craparo, Joe and Sherry Kittinger and supported by Tom Crouch of the Smithsonian Institute, but representing only himself as an historian and scholar, made a visit to the United States Department of the Interior, National Park Service, to encourage establishing the Stratobowl as a National Landmark site.

Extensive and thorough research by the National Park Service prepared the way for this worthwhile designation. The research and communication was handled by Historian Caridad de la Vega, for the National Historic Landmarks Program. However, the National Parks honors the wishes of the landowners and the designation was turned down by some of the private land owners of the bowl.

At the present time some of the Stratobowl private landowners are still considering the possibility of the designation making the bowl a National Historic Landmark.

The overlook on the rim of the Stratobowl with the four monument's testimony, continues to be a favorite 20 minute walk and destination for local and tourist hikers. It is and will be truly a "lasting legacy" for the generations to come.

Chapter Ten

South Dakota Bombed!

*"God would never allow that such a machine be built...
because everybody realizes that no city would be safe
from raids...bombs could be hurled from great heights."*
— Francesco de Terzi

Americans were in shock following the Pearl Harbor attack, December 7th, 1941. On the west coast, blackouts were ordered the very first night for fear of immediate air raids. People were told to place buckets of sand in their homes, with which to smother fires caused by incendiary bombs. [1]

When I was around five years old, I remember my mother practiced putting black shades on the west windows of our Day County, South Dakota farmhouse. It was part of a prescribed drill. During World War II we were instructed by the War Department to know how to cover any light that could be seen at night, should Japanese bombers fly over and pick a lonely farmhouse light as a convenient target.

This precaution, though really unnecessary, we adhered to less out of fear and more motivated by patriotism and vigilance. The cost and sacrifice of the war struck our family more deeply than most. My three cousin – all brothers living 3 miles from our farm, 8 ½ miles west of Butler, in Day County, South Dakota, were all killed in the War.

Radioman 2nd Class Ordien Herr, Radioman 2nd Class Eugene Herr and First Lieutenant LeRoy Herr – *Photos courtesy of Susan Herr Hanson and Elaine Herr*

My first cousin killed, son of Uncle Albert and Aunt Marmien Herr, was Navy Radioman 2nd Class Ordien F. Herr, June 8, 1943, when his plane crashed in the Pacific. Ordien had entered the service 12 days after Pearl Harbor, December 19, 1941. He served in the Pacific Theatre and was wounded. He recovered from his wounds and was transferred to a torpedo bombing squadron. Ordien was en route to Australia for R and R when his place crashed.

My second cousin killed, was Navy RM 2nd Class Eugene Lamont Herr. Eugene had written home October 4, 1944 that his ship was in port for repairs. On October 9, he wrote that he was still well. Eugene was serving in the Pacific Theatre on the Destroyer *USS Johnston* when it was struck by the enemy in the Leyte Invasion in the Philippines. The Battle of Leyte Gulf, Philippines, lasted from October 22-27, 1944. The battle was the most immense sea/air action ever fought in the history of warfare. 166 US and 65 Japanese warships were engaged, plus 1,280 US and 716 Japanese aircraft.

Eugene died October 25, 1944 according to the Bureau of Naval Personnel in a letter to his parents.

> "The *Johnston* was under attack by a vastly superior force of Japanese battleships, cruisers and destroyers. She fought back valiantly for three hours, inflicting heavy damage upon the enemy, before we

were forced to abandon ship. Eugene was uninjured at the time we abandoned ship. However he was suffering from battle fatigue after the strenuous three hour struggle. He was in the water with a life jacket on and clinging to a raft for two days. On the second night Eugene died of exhaustion and exposure, and was buried at sea with a very simple ceremony."

My third cousin killed, was US Naval Air Force Lieutenant LeRoy M. Herr. He enlisted in the US Naval Air Force June 10, 1942. After training at Wold-Chamberlain Airfield in Minneapolis, Minnesota and in Corpus Christi, Texas, Lieutenant Herr served thirty-eight months as a pilot, eighteen of which were in the South Pacific, Fiji Islands.

While flying over Catalina Island on September 26, 1945, LeRoy was killed in a midair airplane crash. LeRoy had a mere twelve hours of flight time left to serve before being discharged from the Navy. LeRoy had married Virginia Byers on December 31, 1944. On December 12, 1945, after ReRoy's death, Virginia gave birth to their daughter, Royalee.

LeRoy was the third son of Marmien and Albert Herr to die within a two and a half year time span in service for their country.

I remember when LeRoy died, and the dreaded telegram had been delivered. My family drove over to the Herr farm to console Aunt Marmien and Uncle Albert. I still remember as an eight year old how Aunt Marmien, who was a wonderful but stoic Norwegian, hid behind the living room door so that we would not see her cry.

I remember the three blue stars on the white banner that hung near the chancel in our church, Bethesda Lutheran in Bristol, and how each star was changed to gold when one of the Herr boys died.

Balloon Attack

During this time, my family and the general populace in American were unaware that the Japanese were plotting and implementing a sinister and surprising attack on the North American mainland, including South Dakota. Even though the war struck our family unmercifully, the battle ground itself seemed far distant. But no longer would that be really true.

"From widespread reports during 1944-1945, along with paper and metal fragments, United States military authorities began piecing together a fantastic story. The following are four isolated, but typical incidents that lead the military to become concerned:

A father and son on an early morning fishing trip were just settling down when they observed a parachute or balloon-like object drift silently by and over the nearby hill. Moments later an explosion echoed through the valley leaving only a small trace of smoke coming from the direction in which the object had disappeared. By the time the two reached the area of the incident, fragments of paper were the only things unusual in the silence of the north woods.

The attention of two farmers at work in their field was diverted by a sudden explosion in a nearby field and an eruption of a cloud of yellow smoke. They moved cautiously to the scene, only to find a small hole in the ground with metal fragments nearby. There was no evidence as to how this mysterious object was delivered.

A mother tucking her sleeping child in for the night was shocked by a sudden flash of light through the window, followed instantly by the sharp crack of an explosion in the silent darkness.

Ranchers coming over the top of a hill near where they had camped the night before, discovered a partially inflated balloon entangled in the scrub brush. It had no bombs, but somewhere along its journey, had discharged its lethal cargo." [2]

It became clear, that the Japanese invasion, which included South Dakota, had began in earnest, when "on Friday, 20, March, 1945 at 6:50 p.m., mountain war time, a large balloon descended toward the Cheyenne Indian Reservation in South Dakota. The bag was about thirty-two feet in diameter and made of smooth pliable paper. A metal gas relief valve covered a hole at the bottom from which nineteen forty-foot shrouds connected the envelope with a mass of ballast gear. The silvery sphere, blown gently by a slight northeasterly breeze, landed in tall grass and bounced along until the equipment caught in a washout. What they found puzzled them. They had never seen anything quite like it before. After considerable discussion, they decided it was a weather balloon of no great importance. Determining that the balloon could still float, they grabbed the shrouds and led the entire contraption back to the ranch. There, firmly tied to a fence post, the bag swayed gently through the night hours." 3

There were several balloon incidents that occurred in South Dakota during the first half of 1945. Other findings appeared near Buffalo, Kadoka,

Marcus, Wolsey Red Elm and Madison. A balloon exploded in the middle of March in broad daylight in the sky north of Custer. The Custer incident was witnessed by many, while another balloon sighting was witnessed a few days later. A balloon appeared over Belle Fourche one afternoon at about 3 to 4 thousand feet high. A civilian pilot who pursued the object reported: "Catching the late afternoon rays of the sun, the balloon appeared in the sky as a perfect silvery sphere which could be seen only if the observer was in line with the reflection. At times, it disappeared in the blue haze and near Piedmont a squadron of flying fortresses from the local air base passed within a quarter of a mile without noticing it." [4]

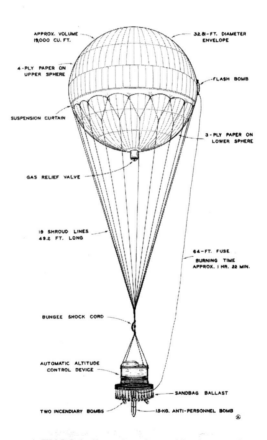

A FUGO balloon bomb used in the attack
on South Dakota and North America.
— *Figure #39 reprinted from Mikesh (1973) with permission
of Smithsonian Institution Scholarly Press.*

"Balloons continued to drift over South Dakota – one South Dakotan unknowingly carried a balloon bomb many miles over bumpy roads in the trunk of his car, and another South Dakotan allowed his children to use a balloon bag for a doll house when in reality the balloon bombs were quite deadly. The charge from one balloon, exploded with a dynamite cap by an army intelligence officer at Rapid City. It tore a hole in the ground three feet deep and five feet in diameter. Of course, this was what the balloons were designed to do...as military weapons." [5]

A Fugo balloon was discovered on the ranch of Mrs. Ray Waring, 12-14 miles south and one mile east of Ree Heights. The device consisted of a barometer, battery, wiring, blow-out plugs and sandbags. When the balloon fell below 3,000 feet while crossing the Pacific, the barometer would activate the battery, setting off a blow-out plug that would in turn release a sandbag, thus enabling the balloon to ascend. After all the sandbags were released, the bomb was dropped.

Mrs. Waring kept the balloon and ballast dropping device for about a month. She first called the sheriff, but when he did nothing about it, she called Les Price of the Attorney General's Office, who came and got it. The Warings then turned the find over to the state law enforcement officers from the State capitol. It now resides in the State Historical museum in Pierre, South Dakota. Mrs. Ray Waring gave this report and information to Leo O'Neal on the 14th of May, 1972.

One of the balloon bombs that was found on South Dakota soil.
— Courtesy of the South Dakota State Historical Society Museum

119

Of 308 balloon incident findings chronicled from November 1944 to August 31, 1945 by Bert Webber, the following balloons and/or fragments were found at the following sites:

"Nowlin/Rapid City	February 12
Buffalo	March 6
Ree Heights/Wolsey	March 22
Kadoka	March 26
Red Elm	March 30
Marcus	March 31
Wolsey	April 13
Buffalo	April 26
Madison	May 26" 6

Japanese FUGO Balloon Bombs, undiscovered, still exist today and threaten public safety. One example of a later find, was in 1983 by Thomas Newcomb near Lake of the Woods in Oregon.

Three South Dakota relatives, Carl Baldwin, 51, his brother, Brad, and his sister Vanessa Boy, lost an aunt in the only balloon bomb explosion that took American lives on the North American mainland. The tragedy took place on Gearhart Mountain near Bly, Oregon, May 5, 1945. It happened on a Saturday morning when Archie Mitchell, his wife Elsye, and five Sunday School students, Edward Engen,13, Jay Gifford, 13, Sherman Shoemaker, 11, Joan Patzke, 13, and Dick Patzke, 14, went on a picnic. One of the children nonchalantly kicked at a piece of metal in the underbrush. It was a Japanese incendiary bomb hidden in the forest. There was a flash of fire! A blinding light! Shrapnel ripped through brush and human limbs, and six people were killed. Archie Mitchell, pastor of the Christian and Missionary Alliance church, was a distance away and was uninjured in the blast.

In an interview by Steve Young of the Argus Leader, May 2, 2002, prior to the 57th anniversary of the explosion and deaths, (May 5, 1945), the two nephews and one niece from Sioux Falls lamented the tragedy. Carl and Brad Baldwin, and sister Vanessa Boy, mourned the death of their Aunt Elsye Mitchell who they never got to meet. "She died before any of us were even born. It's been a sad deal, especially for our mother." Indeed, it's likely that their mother Eva Fowler, Elsye's sister, will make her way again this month to the mountain country northwest of the small Oregon logging community of Bly. There she will visit the stone monument erected August 20, 1950 by the Weyerhaeuser Timber Company on the site of the explosion." (Steve Young Interview)

Balloons in Warfare

Balloons used in warfare, prior to the bombing of South Dakota and the rest of the North American mainland, was as old as the invention of the balloon itself.

After the Montgolfieres and Charlieres flights, old Benjamin Franklin wrote to a friend, Dr. John Ingen-Hausz, "Convincing sovereigns of the folly of war may perhaps be one effect of it, since it will be impractical for the most potent of them to guard his dominions. Five thousand balloons, capable of raising two men each, could cost no more than five ships of the line, and where is the prince who can afford so to cover his country with troops for its defense as that ten thousand men descending from the clouds might not in many places do an infinite deal of mischief, before a force could be brought together to repel them?" (Source: Ben Franklin correspondences)

Thus the American statesman accurately predicted the use of airborne armies during war.

"The use of hot air balloons in warfare was not new to humankind. Ever since November 21, 1783, when the Frenchman Jean Francois Pilatre de Rozier became the first man to fly in a balloon over Paris, nations recognized the advantage of air reconnaissance of their enemy in the time of war. The outbreak of the French Revolution in 1789, followed by a long war on a large scale, paved the way for balloons in military operation. Napoleon's 1812 campaign included balloons, as well as did the later campaign for Italian independence from Austria in 1859. And in America, the idea was germinating as well. The Seminole War in Florida in 1835, over the removal of the Seminole Indians to the Western U. S. reservations, dragged on for several years. In 1840, Colonel John H. Sherburne suggested to Secretary of War, Joel Poinsett, that balloons might be assigned for service. The war however ended before any balloon plans were implemented.

During the Mexican War of 1846, John Wise, the widely known aeronaut from Lancaster, Pennsylvania, proposed to the government a strategy to clear the way for the capture of Vera Cruz, Mexico. Vera Cruz, which was under siege, could be defeated by means of percussion torpedoes dropped from captive balloons proposed Wise. The plan never got off the ground, so to speak." [7]

The first military use of a hot air balloon happened during the Battle of Fleurus (1794) where the French used the balloon *l'Entreprenant* as an observation post.

In the United States, during the Civil War, balloons were employed by both the North and the South. Key figures were James Allen, John Wise, the veteran balloonist, Pennsylvania aeronaut, John La Mountain and Thaddeus Sobieski Constantine Lowe, of the Federal army and to a lesser degree, George Armstrong Custer, and others. War balloons were used primarily as observation posts.

The Franco-Prussian War of 1870-71 saw Paris completely cut off from the outside world by Bismarck's Prussian Army. "The balloon's proudest hour came during the Seige of Paris in the Franco-Prussian War. Exactly sixty-eight balloons left the city during the blockade. They carried 400 pounds of mail each, a lot of carrier pigeons, and a very inexperienced collection of pilots. Only five balloonists were in Paris when the siege began, and so most of the escapees were having their maiden flights. A few landed in German-occupied territory, but most got to safety....the pigeons which survived the flight were released to take back news to Paris. On their legs they carried micro-film of newspapers and letters. These were projected on to screens, copied and then reprinted and distributed throughout the beleaguered city. The balloons could fly at a height where they were safe from the Prussian bullets, but the pigeons flew much lower and many became pie for the Uhlan's pots." [8]

Besides mail, some paying passengers were able to escape, although one passenger got farther than he had bargained for–the wind carried him all the way to Norway.

Japan was influenced by French sporting balloons until France lost the Franco-Prussian War. After that, Japan followed the German examples, studying Kaiser's balloon technology, and eventually creating a military balloon corps based on the Prussian model. As a result, Japan's military balloonists were ready for action when their country went to war with

Russia in 1904. Later, after further development, Japan, stirred with fury, needed to avenge the United State's Doolittle Raid on their homeland. Japan created and launched the attack by balloon, on South Dakota and the rest of the American homeland.

Balloons and Propaganda

One unholy marriage united propaganda and ballooning. "Japanese propaganda broadcasts mentioned great fires and an American populace in panic. One broadcast said that five hundred casualties had occurred. Another broadcast raised the figure to ten thousand. Several million airborne troops were said to be ready to follow the balloons. As it turned out Elyse Mitchell and five Sunday School children near Bly, Oregon, were the only actual casualties." [9]

There was no invasion and no panic. South Dakota newspapers following the war department's mandate, wisely maintained self-censorship when a report surfaced about a mysterious Japanese Fugo balloon sighting.

Germany also used propaganda balloons during World War II. The South Dakota State Museum in Pierre, received from Evangeline Rahn, a former student in the Pierre Schools, one of the little silk balloons used by the Germans for spreading propaganda within the American lines. It was given to her by brother Perry who served as a soldier. The message began, *"How to end the war. Do your part to end the war–stop fighting.... the tales they tell you of the cruelties of the German prison camps are fairy tales. Of course you may not like being a prisoner of war...wake up and stop the war...."*

Later, the United States would use propaganda balloons during the Cold War to urge citizens in captive countries to seek freedom.

In the late 1950's, South Dakota's Ed Yost, when working for General Mills in Minneapolis, developed balloons carrying propaganda leaflets that flew over the Iron Curtain. General Mills, sub-contracted by Radio Free Europe, distributed propaganda leaflets behind the Iron Curtain from bases in West Germany. Yost and team designed and built the propaganda balloons, each of which lifted off from the ashes of the war-ravaged, country like the mythical Phoenix. General Mills Inc., Mechanical Division, announced to it's employees on August 14, 1951: *"Tens of thousands of General Mills-made balloons are now landing in Czechoslovakia and Poland, carrying messages of hope to peoples behind the Iron Curtain. Called pillow balloons because of their 54' square size, they were developed at company research laboratories in 1949. The balloons made of polyethylene, a substance commonly used in food saver bags. The company is one of two manufacturers making balloons for the Crusade for Freedom, National Committee for Free Europe, sponsors of this project."* [10]

Was the leaflet propaganda by balloon an effective campaign? Ed Yost, when asked this question, responded: "The thing worked too 'darn' good and we got the Hungarian Revolution."

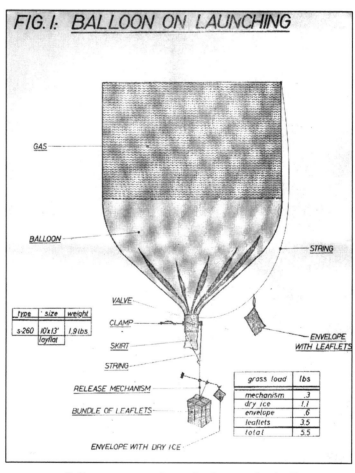

Balloon propaganda mechanism used to leaflet
Communist occupied countries. – *Courtesy of Ed Yost*

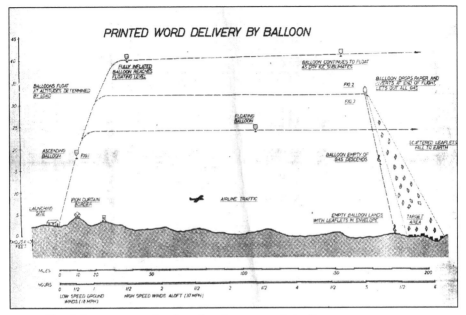

The delivery system for the printed propaganda leaflets – *Courtesy of Ed Yost*

Balloons for Freedom

In 1959, Ed Yost, then Raven Industries Operations Manager in Sioux Falls, announced a gas balloon flight that would begin on a future Saturday night that would carry him over Dubuque, Iowa, Chicago, Pittsburgh and finally land in the New York area the following Sunday evening. This flight would be called the *Flight for Freedom* and sponsored by the Crusade for Freedom national office. The flight was planned to raise $10,000 . The flight would fly from Sioux Falls, South Dakota to the East Coast in conjunction with a local radio marathon to raise the funds. Raven Industries donated the balloon, controls and instruments while Yost donated his time and expertise.

"On Saturday evening, March 7, 1959, the weather was less than perfect for a balloon flight. But the radio marathon had begun and Yost decided to continue with the balloon flight. At Joe Foss field in Sioux Falls, searchlights shone, sirens blared and a large crowd estimated to be over 1,000 onlookers, cheered as Yost, wearing three suits of underwear, fur-lined boots and nylon coveralls, lifted off at 6:30 p.m. in the 50 x

25 foot balloon. State and national flags were suspended underneath the gondola and inverted tear drop envelope. Yost waved as the gondola lifted. Aboard was a parachute, oxygen, Raven radio equipment, hot coffee, 15 sandwiches and a red aircraft warning light.

Yost radioed 40 minutes later that the balloon had reached 8,000 feet. He was traveling at 30 miles per hour. After crossing the Minnesota line the weather turned to freezing rain and snow, forcing Yost to land at the Herbert Larson farm near Amboy, Minnesota. Ed Yost commenting later said, 'I could see the light on at the Larson place, although it was after midnight. It looked like a nice place so I decided to put down there. The Larsons were a bit surprised. But they certainly were hospitable. Mrs. Larson fried some eggs and put the coffee pot on the stove. The bed they fixed for me was a lot more comfortable than the four by five, one-foot deep gondola I was riding.'

Even though the flight was not successful in reaching it's announced final goal of landing on the east coast, the Crusade for Freedom benefited from newspaper coverage and garnered much publicity for the cause." 11

Balloons in history have been used for bombs, propaganda and freedom.

During the Cold War era, balloons were used not only for propaganda, but as actual vehicles to achieve freedom. After the hot air balloon had been invented by Ed Yost, the phenomenon stirred the imagination and courage of folks in Communist countries. Two courageous examples are remembered:

On September 16, 1979, Peter Strelzyk, a 37 year old electrician with some earlier experience in the East German Air Force as an aircraft mechanic, piloted a homemade balloon with wife, Doris, two children and good friends Gunter Wetzel, his wife, and four children, to a height of about 8,000 feet. They quietly passed over sentries, attack dogs, watchtowers, electric fences, land mines and remotely triggered shrapnel guns mounted at leg, midsection and head levels, down to safety and freedom in the town of Naila, West Germany. This was the second try and thankfully a successful escape attempt. In 1982, Disney film chronicled this amazing flight to freedom in the movie *Night Crossing*.

A second balloon escape from the Iron Curtain Country, Czechoslovakia, was attempted by thirty-six year-old Robert Hutyra, his wife and two children. The Hutyra balloon was similar in design to the

Peter Strelzyk balloon, having a wooden platform, a propane burner, and little else. The family landed close to the town of Drasenhofen in Eastern Austria, and were able to walk into town where police were notified of their arrival. The balloon's envelope had been sewn secretly in the Hutyra home and was launched from a desolate forest near the Austrian border.

Though the balloon was often used as an instrument of war, such as the bombing of South Dakota and the North American homeland, there are many examples in which the balloon served to free people's physical lives, as well as their minds, and spirits from the bonds of oppression.

Chapter Eleven

Excelsior (Ever Upward)

"I take a breath and hold it. This is no fantasy. I am really here....Lord take care of me now...I jump from space."
— Joe Kittinger

"Lord, Take Care of me now," Captain Joseph Kittinger uttered as he leaped from an open gondola 18 ½ miles above the earth. The helium balloon had taken the aeronaut to a record height of 102,800 feet.

For the next four minutes and 38 seconds Captain Joe plummeted in his pressure-suited outer layer at speeds up to 614 miles per hour, on the verge of mach one.

"I free-fell for 16 seconds before I felt a soft shudder on my back as the pilot chute popped out followed by a 5 foot stabilization chute right on cue."[1] At 17,000 feet the main chute opened automatically and 9 minutes later, Joe landed safely on the New Mexico bed of grass, sand

Joe Kittinger leaping at 102,800 feet and parachuting safely to earth
— Photo used by permission fromVolkmar K. Wentzel/National Geographic Stock

and sage. "Ahhh boy! Lord thank you for protecting me during the long fall." [2]

Joe Kittinger set the world's record for highest ascent and longest parachute jump on that momentous day – August 16, 1960. The record stood for 52 years till this past October 14, 2012.

Joe Kittinger and South Dakota's Ed Yost might well be called the "dynamic duo" in the science, sport and history of ballooning. Joe and Ed were close friends. Joe was a supporter and friend of several significant ballooning events in South Dakota. In 1960, at the 25th anniversary celebration of *Explorer II*'s record flight from the Stratobowl in 1935, Captain Joe was the featured speaker. The celebration took place at the Stratobowl rim near Rapid City. At that same time Captain Joe was being honored as one of several special guests at the *Explorer II* celebration, Ed Yost, Jim Winker, and the Raven Industries team from Sioux Falls, were busy launching the 2nd first experimental flight of a modern hot air balloon, below, in the Stratobowl.

Years later, in 1984, retired Colonel Joe, would fly the first solo across the Atlantic in a Yost-built helium balloon named the *Balloon of Peace*, constructed at Tea, South Dakota.

Joe had met Ed earlier during Project *Manhigh*, when Ed was working for the General Mills Balloon Division in Minneapolis. Joe recalled, "later while Ed was working at Raven Industries on a contract with the Office of Naval Research in 1960—-just a couple of months after my flight in *Excelsior III*—he had invented, built and flown the first modern hot-air balloon. His break-through design consisted of a nylon envelope, a propane burner, and a half-inch piece of plywood for a seat. On that maiden flight, Ed Yost lifted off from a field in Bruning, Nebraska, and stayed aloft for twenty-five minutes. Later, Ed experimented again in the Stratobowl, as mentioned above, and this time stayed up for two hours and reached an altitude of 9,000 feet.... In spite of a lack of much formal education, he was the closest thing to a mechanical genius I'd ever met. He could design anything, build anything, fix anything. He'd single-handedly launched the era of modern hot-air ballooning." [3]

Joseph Kittinger Jr. was born in Tampa, Florida on July, 1928, and after three months, he and his family moved to Orlando. Joe raced hydroplanes as a young man but was mostly fascinated by airplanes such as the Ford Tri-motor he spied at a nearby airport. As a young 17 year old kid, he soloed in a Piper Cub. Kittinger attended the University of Florida and then left to join the U. S. Air Force in 1949 as an aviation cadet. His love was fighter planes. He served as a NATO test pilot in Germany until 1953. He was then assigned to the Air Force Missile Development Center at Holloman Air Force Base in New Mexico. Captain Kittinger would eventually, throughout his career, fly 93 different air crafts.

When the UFO craze, during the late 1940's and 1950's was all the rage, Joe tells in his autobiography a fascinating story. "Balloons were always being mistaken for UFO's in the 1950's – especially at twilight when a balloon at altitude was catching the sun and glowing even after darkness had fallen for observers on the ground." [4]

One particular incident caused more than its fair share of controversy. "In May, 1959 Joe was flying a balloon training flight with Captain Dan Flugham and Captain Bill Kaufman out of the Holloman Air Force Base in New Mexico when they landed very hard near Roswell, New Mexico. Dan got hurt with burst blood vessels in his head. His head swelled up like a basketball. Somebody at Roswell saw this 'creature' with a huge grotesque head being raced away (to another hospital). This incident became a part of the ongoing Roswell legend that had its genesis in the summer of 1947 when a rancher reported evidence of an alien landing in the desert. The alien debris turned out to be part of a classified Air Force high altitude balloon program called *Project Mogul*." [5]

In 1955 Joe Kittinger flew the T-33 observation plane that monitored the "rocket-sled" experiment of aircraft medicine pioneer, Colonel John Paul Stapp. Stapp took his aircraft sled to 632 miles per hour (1,017 kilometers per hour) to test how gravitational stress affected the human body.

Joe worked closely with Colonel Stapp during the stratospheric balloon flights and touted his profound respect for Stapp all his life.

Stapp recruited Kittinger for Project *Manhigh*. This project, begun in 1955, would utilize balloons capable of high altitude flight, lifting a pressurized gondola in order to study cosmic rays and to determine if humans could fly safely at an altitude above 99 percent of the Earth's atmosphere.

Kittinger made the first *Manhigh* ascent on June 2, 1957. This was launched from South Saint Paul, Minnesota, and attained an altitude of 96,000 feet. Joe said he saluted aeronauts Stevens and Anderson as he passed their record flight of 72,395 feet. *Manhigh II* was flown by Major

David Simons which reached an altitude of 101,500. *Manhigh III* was flown by Lieutenant Clifton "Demi" McClure and reached an altitude of 99,700 feet. 6

In 1958, Kittinger transferred to the Escape Section of the Aeromedical Laboratory at Wright Air Development Center's Aero Medical Laboratory. November 16, 1959, Kittinger piloted *Excelsior I* to 76,000 feet. He nearly lost his life but landed safely. *Excelsior II* climbed to 74,700 feet and then on August 16, 1960 he set the record jump at 102,800 feet.

Later, Kittinger volunteered for three combat tours in Vietnam, flying 483 missions. He shot down one MiG hoping to get two more and achieve the title "ace." Kittinger relished the fact that his life and adventures were "lucky," although on May 11, 1972, his "luck" ran out as he was shot down just four days before he was scheduled to go home. He spent 11 month in the infamous Hanoi Hilton where he was tortured as a captive prisoner of war. He told this author how they fed him watery pumpkin soup. This was one of many numerous indignities his captors inflicted on him and the other prisoners. This behavior was and is contrary to the accords of the Geneva Convention. Prisoners are to tell only name, rank and serial number. Order and rank were maintained, even in a prisoner of war setting. Joe, as the Senior Ranking Officer, provided exceptional leadership, while keeping morale high as possible among his fellow prisoners.

During his captivity, Joe maintained sanity by dreaming of a balloon flight around the world. Often stuck in solitary confinement, Joe mapped out the balloon flight in detail. By retaining purpose and meaning, despite his dire circumstances, he was able to remain strong. "I thought back on my balloon training, on the *Manhigh* and *Excelsio*r and *Stargazer* days. I began to think about what it would take to make a solo flight around the world in a balloon. It was kind of a romantic, crazy idea—but what a glorious trip that would be! It was one of the last great adventure challenges out there, and I saw no reason why I shouldn't be the one to do it. There were a million things to consider with such a flight, and I had time to consider them all. I came up with a balloon design that would take advantage of solar heating and nighttime cooling, the gondola setup I'd need, the entire life-support and communications system, and on and on. This fantasy project kept me alert and mentally occupied. As the days and weeks ground on, I continued to review and modify these plans until

I begin to believe that they *would* work—just as soon as I could find my way out of this hellhole." [7]

According to a nation-wide Gallup poll in the 90's, the primary human Spiritual Need in America was the need to maintain meaning and purpose in life. In normal circumstances, meaning and purpose is a challenge but to maintain meaning and purpose in challenging circumstances is even more crucial.

Viktor E. Frankl, a concentration camp survivor and psychotherapist who wrote *Man's Search for Meaning,* in his recollections of imprisonment in death camps during the Holocaust, observed the difference between those who endured and those who died. The difference he described as follows: "One of the prisoners, who on his arrival marched with a long column of new inmates from the station to the camp, told me later that he had felt as though he were marching at his own funeral. He regarded it as over and done, as if he had already died....A man who let himself decline because he could not see any future goal, found himself occupied with retrospective thoughts...the prisoner who had lost faith in the future—-his future——was doomed. With his loss of belief in the future, he also lost his spiritual hold; he let himself decline and became subject to mental and physical decay." [9]

A fellow prisoner for eight years during the Vietnam War was Admiral James Stockdale. When Stockdale was asked, "Who didn't make it out?" Admiral Stockdale replied, "Oh that's easy. The optimists...they were the ones who said, 'We're going to be out by Christmas.' And Christmas would come, and Christmas would go. Then they'd say, 'we're going to be out by Easter.' And Easter would come, and Easter would go. And then Thanksgiving, and then it would be Christmas again. And they died of a broken heart." [10]

Prisoner of war, Lieutenant Colonel Joseph Kittinger, focused on the future and the adventures the future would bring. When released, one of his ballooning dreams (see chapter 16, Crossing the Pond) would actually be fulfilled, at least in part, in 1984.

Joe retired from the Air Force in 1978 as a colonel. He soon began entering balloon competitions. He won the Gordon Bennett Gas Balloon Race in his balloon the *Rosie O'Grady* four times during the 1980's and was able to retire the Gordon Bennett Gas Balloon trophy after three consecutive victories. When in 1976, Ed Yost called Joe and said he (Ed), was going to try to fly the Atlantic and he wanted Joe to serve as Ed's Chief of Operations. Joe agreed. Ed launched from Milbridge, Maine, but crashed into the Atlantic 700 miles short of Portugal.

In 1982 Joe met Sherry Reed who would become his perfect companion and eventual wife.

In the spring of 1984, Ed Yost contacted Joe saying he had a sponsor for a transatlantic solo flight. After preparations were made, Joe launched in a Yost South Dakota-made balloon, called the *Balloon of Peace,* from Caribou, Maine. The flight was successful after an amazing, challenging four day trip from Caribou, Maine to Cairo Montenotte, Italy. Joe's flight set a record for both the longest solo balloon flight and a distance record for the type of balloon used. His "Hanoi Hilton" determinations served him well.

Joe Kittinger is seen and described as the redneck fighter pilot and world class balloonist. To summarize Joe's accomplishments Glen Moyer described it this way: "In world records you learn that Joe Kittinger has made the world's highest parachute jump....has made the most high altitude balloon flights of anyone in history, that he holds a distance record for AA6 and AA7 gas balloons (2,001 miles in 72 hours), that he has been awarded two Montgolfier Diplomas and finally that he was the first man to fly solo across the Atlantic Ocean in a balloon. Impressive. Every time he steps into — or in some cases out of — an aircraft, he seems to make history. But who is this man in the red bandanna? Another writer once said of Kittinger, 'Roll Chuck Yeager and Evel Knievel into one, and you get an idea of Joe Kittinger's thirst for adventure.' After my brief visit with him, I would add a third name to that list, Indiana Jones. For it seems that no matter what the danger or personal peril facing him, Joe Kittinger always manages to reach into an invisible haversack and pull out a winning hand. No guts, no glory? Joe Kittinger has plenty of both." [12]

Presently, (2012, 2013) Joe and Sherry are barnstorming the country, promoting Joe's autobiography *Come Up and Get Me.* The phrase, and eventual title, *Come Up and Get Me,* was tapped in Morse Code to David Simons and crew while Joe was at peak altitude in *Manhigh I.* Simons feared Joe was going to bail out, thinking he had been seduced by the rapture of space. Joe got the last laugh and an intriguing title for a book.

For the past four years, 2008-2012, Joe had been the Operations Manager for Austrian Alex Baumgartner's attempt to break the highest parachute jump in history. The record jump at 128,000 feet happened on October 14, 2012 near Roswell, New Mexico breaking Joe's own record. See details in Chapter twelve of **Balloons Aloft: Flying South Dakota Skies.**

Chapter Twelve

Strato-Jump: A Courageous Attempt?

"There are no secrets to success. It is the result of preparation, hard work and learning from failure." — Colin Powell

T he book came unannounced, apparently a pre-Christmas, Christmas present from Vadito, New Mexico. The book, masterfully researched and written by Craig Ryan was titled, *The Magnificent Failure*. The hand-written note scrawled on the inside cover and the gifter, I recognized. The message said simply, *"10 December 2003. To our Precious friends —-they create happiness in our lives —-Rev. Arley and Pam Fadness. With smiles and tears of joy Ed Yost & K9 Shadow.*

I thanked Ed and quickly devoured Craig Ryan's *The Magnificent Failure*, fascinated by this "one man's" quest to fly higher and further than any person had done before and then parachute free-fall landing safely on the earth. The man's name — Nick Piantanida.

Before we visit the incredible story and journey of Nick Piantanida in the 1960's, we fast forward to today's similar

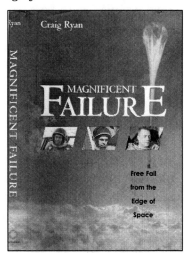

The Magnificent Failure chronicles the amazing story of Nick Piantanida.
– Book jacket by permission of Smithsonian Institution Scholarly Press

135

but <u>successful</u> adventure, 56 years later, that has just unfolded. It was called the *Red Bull Stratos Project.* According to the National Geographic special telecast called *SpaceDive,* the public learned that on March 15, 2012, Austrian skydiver extraordinaire Felix Baumgartner joined an elite company of men who had jumped from the edge of space and landed alive. Felix Baungartner reached 71,520 feet in a pressurized capsule before he stepped out into nothing, and began his 8 minute fall back to earth. He landed safely near Roswell, New Mexico. The 42 year old Felix Baumgartner, had now flown above the "Armstrong Line." (The Armstrong Line or Limit is that extreme altitude beyond which humans could not survive in an unpressurized environment. That altitude is between 62,000–63,500 feet or about 12 miles above the earth).

During the fall, "Fearless Felix" as some referred to him, reached astounding speeds of 364.4 miles per hour – more than 534 feet per second. "The free-fall portion of his dive lasted 3 minutes and 43 seconds, after which he pulled the ripcord on his parachute to make the rest of the journey...Joe Kittinger was on hand for the test jump." [1]

Joe Kittinger who held the current highest jump record at 102,800 feet (See chapter Eleven–*"Excelsior (Ever upward)"*), served as Baumgartner's mentor and Operations Manager. "Joe said before the launch, 'Felix, you're going to have one heck of a view when you step out of that door – enjoy the experience.' That's exactly what happened. 'The view is amazing, way better than I thought,' Baumgartner reported." [2]

The March 12th successful jump readied Baumgartner for the next test launch in which he jumped from 90,000 feet before attempting a third jump which was destined to reach 120,000 feet and thus break Joe Kittinger's record.

I received on September 30, 2012, this email, "Monday, Sherry and I depart for Roswell, NM where we will be preparing for the Project Stratos Jump from above 120,000 feet. As you know we have already successfully completed two manned jumps from 72,000 and 97,000 feet in preparation for the next jump from space. The balloon which has a capacity of 29 million cubic feet is huge, weighing almost 4,000 pounds and standing over 700 feet above the ground on launch. As a result, to inflate this very fragile balloon requires perfect weather and wind conditions. The earliest date that we plan on launching is 8 October just at sunup. The climb to altitude will require 2 hours 15 minutes. Felix should go supersonic after about 30 seconds of free fall. His free fall should last about 5 minutes and 30 seconds before he opens his main parachute about 5,000 feet above the desert. We expect over 300 of the media to attend the launch, including the BBC which is making a documentary of the Project and jumps.

I have worked for over 4 years on this project and have been honored to be a part of this historic occasion. Felix is ready, the team is ready, the capsule is ready – all we need is good weather (and some Divine cooperation) to successfully conclude Project Stratos. Buoyantly, Joe and Sherry Kittinger."

"Baumgartner was well known for his record-setting Base Jumps and Skydives. He set his first world record in 1999, when he made the highest parachute dive from a building from the Petronas Towers in Kuala, Lumpur, Malaysia. In July 2003, he became the first person to cross the English Channel in free fall using a specially made fiber wing. He also set a record for the lowest Base Jump ever, and was the first person to Base Jump from the Milau Viaduct in France in June 2004.

In December, 2007, he became the first person to jump from the 91st floor observation deck of the then tallest completed building in the world, Taipei 101, Taipei, Taiwan." [3]

The goal of the Red Bull Stratos Project was to free-fall from 120,000 feet, thereby enabling the first man to go supersonic without the support of a vehicle.

There were no hard feelings between Joe Kittinger and Felix Baumgartner. Baumgartner was convinced he could break the long-standing record of 102,800 feet owned and celebrated by Colonel Joseph Kittinger.

We go back now, to the early 1960's, to make the "South Dakota Connection." We met an ambitious, determined, gutsy, Nick Piantanida, who would do whatever it takes to break that 102,800 high parachute jump held by Joe Kittinger, and just as importantly, the longest free-fall record held by Eugene Andreev.

By the spring of 1961, manned high flights had virtually come to an end. The highest manned flight belonged to Commander Malcolm Ross and Lieutenant Commander Victor Prather of the U. S. Navy. They ascended to 113,740 feet, but ended tragically, when Victor Prather drowned, while being rescued in the splash down.

While all eyes were upon astronauts preparing to fly to the moon — on the second day of February, 1966, Nick Piantanida flew a full two miles beyond Ross and Prather. And he did it mostly on his own. He raised his own funds and planned his own project. He recruited his own staff and – both for better and for worse – called his own shots, and yet by the spring of that same year both fate and the times would catch up to Nick Piantanida. Nick came up short of his ultimate goal—to set a new world free-fall record. His efforts are judged a failure. But were they?

Nick Piantanida's *Strato-Jump I* was launched at the old University of Minnesota Airport off Highway 8 in the town of New Brighton, just five miles northwest of St. Paul. The Litton built balloon carried Nick to 23,000 feet before it malfunctioned, and the flight *Strato-Jump I* had to be terminated. Nick parachuted to safety, landing in a city dump known as the Pig's Eye Dump. "Not very glamorous," Nick would say with a resigned shrug, "but the dump was the safest place to land." [4]

Nick moved quickly and decisively to South Dakota, contacting key players at Raven Industries in Sioux Falls. With the Raven team, he planned his second jump – *Strato-Jump II*.

"One of the first members of the Raven family Nick got to know was co-founder, Paul 'Ed' Yost, a stocky, salt-of-the-earth Midwesterner with a brusque manner and an irreverent hands-on approach to aeronautical engineering. Yost was a blue-jeans, beer-drinking guy with a broad face, an imperial Roman nose, and a shock of wind-fanned hair. He was a gifted storyteller with a rich sense of humor—one of the *Strato-Jump* crew members described him as a 'Will Rogers character'—but he was never the easiest man in the world to work with. His instinctive style of problem-solving often clashed with the by-the-book methodologies of his engineering colleagues, and he could occasionally be stubborn and dictatorial. On the other hand, he was indisputably one of the best scientific balloon flight operations manager around..." [5] Ed, as Nick's flight manager, would serve Nick well, even though Ed had certain reservations.

Strato-Jump II was on the way to a perfect success, reaching 123,500 feet, but then a problem arose. Nick was unable to disconnect his oxygen hose – it had frozen shut. The flight was terminated and coached down by Ed Yost, Dick Keuser and crew. "All I needed was a dollar and a quarter crescent wrench," Nick muttered when he landed safely between the towns of Elmore, Minnesota and Lakota, Iowa. The February 3rd, 1966 Argus Leader headlined, "Piantanida to Make new Free-Fall Try."

"'I fell to 97,000 feet before the chute opened. I had visions of being jerked out of the gondola. But this wasn't the day the man up there wanted me.' Piantanida, 33 year-old parachute instructor from Brick Town, N.J., said he felt the period of free fall by the gondola, confirmed his theory that no stabilizing device is needed in a high drop." [6]

"Meanwhile, in a bizarre footnote to the landmark flight's story, a local farmer who had stood in rapt astonishment as he watched the curious silent aircraft with its space-suited humanoid appear out of nowhere, staggered and dropped to the dirt with a heart attack." [7]

After *Strato-Jump II*, Nick would be grounded – but not for long! The quest to penetrate the stars appeared to be burned into the very DNA of the human specie. Among the trail-blazer daredevils to set high altitude world record jumps were:

Victor Hermon, in 1934, at 24,000 feet.

Art Starnes, in 1941, at 38,800 feet.

Joesph Kittinger, in 1960, at 102,800 feet.

Eugene Andreev, in 1962, at 83,523 (Free fall).

The Victor Herman story unfolds into an unbelievable tale. Victor was an American living in Russia because his father worked for the Ford Motor Company building truck plants during the time of the Stalin purges. Victor loved flying and soon got into parachuting when he found a parachute among some abandoned airplanes in a hanger. He jumped without a reserve parachute. "One day the Russians asked if Victor could jump with a dog. They provided him with a smaller parachute for this purpose. He kept this parachute and used it for his reserve afterwards." [8] On September 6, 1934, after stripping an ANT-9 airplane of all extra equipment in order to attain a height of 24,000 feet (he even threw away his reserve parachute to lighten the load), Victor disconnected his oxygen hose and did a free fall, making his 43rd jump. He did a "dead fall" delay, as it was called in those days, until he was 1500 feet above the earth.

The Russians hailed Victor a hero. Victor was born in Detroit, Michigan and when the Russian officials insisted he renounce his citizenship, he refused. He did a few more jumps and since he made the decision to keep his American citizenship, he was eventually arrested in 1938. Many of the Ford Motor Company employees, where his father worked, were also arrested as spies under the heavy dictatorial hand of Josef Stalin. "Victor was beaten and tortured. After a month of enduring treatment he signed a 'confession'. He was sentenced to death. After a month on death row, his sentence was reduced to 10 years in an extermination camp. He was exiled to Siberia in 1948 for the rest of his life. Victor married a Russian gymnast. The couple lived in exile in a cave chopped out under the ice. Unexpectedly, the Russian officials reversed their sentence and exonerated Victor in 1956. Victor spent the next 20 years trying to leave Russia and return to the US. He finally made it to the US in 1976.

Victor wrote the book *Coming Out of the Ice* to describe his ordeal. The book was made into a movie and is also on audio tapes, read by himself. He died at the age of 69 in 1985, on US soil." [9]

Arthur H. Starnes set the high altitude jump record at 38,000 feet in 1941. Art Starnes was a barnstormer and one of America's ace stunt men. He was a stuntman with the Roscoe Turner Flying Circus. He made the first authenticated high altitude low-opening HALO controlled parachute free fall with adaptive equipment from 30,800 feet – almost six miles–before opening his parachute at 1,500 feet above the ground. He proved that aviators could survive extreme delayed-opening ejections from disabled aircraft.

"Art Starnes made his first parachute jump in the mid-twenties, at the age of 18, with a barnstorming troupe at an airfield outside Charleston, West Virginia. As parachutes go, Art's was, to say the least, basic. Instead of having a harness, there was a large rope loop with a single piece of garden hose covering it. Starnes 'Showman, Aerial Maniac' was not calculated to inspire confidence. In truth, he was a meticulous pilot, famous among aviators for his relentless preparations and rehearsals. It was Starnes, who first developed the technique for free-falling. Regular parachutists viewed the prospect of free-fall with terror. Free-falling meant uncontrolled tumbling and spinning and certain death. Military pilots were taught to pull the ripcord the moment they left their plane. In war, though, this left them exposed to enemy fire. No one yet knew what position they should adopt to stay stable as they fell in the air. Starnes thugged out the answer by trial and error, enduring any number of rolls, tumbles and flat spins before he hit upon the ideal skydiving posture; spread eagle, chest out, knees bent. By the early 30's, Starnes was jumping out of planes, and falling for three and a half miles before pulling his cord." [10]

Nick Piantanida, preparing to set the record for free-falling from a height higher than Joe Kittinger's, was acutely aware of the flight and jumps of both Russian Aeronauts — Pyotr Dolgov and Eugene Andreev. It was in the Spring of 1962 that Colonel P. I. Dolgov and Major E. N. Andreev were appointed to test new high-altitude equipment in the extreme conditions of the stratosphere.

On November 1, 1962, Dolgov and Andreev were lifted into the stratosphere in *The Volga* balloon and gondola in order to test an experimental pressure suit. Eugene Andreev jumped safely. He landing near the city of Saratov, and holds the official FAI record for the longest free-fall parachute jump at 80,380 feet. This record setting flight was on November 1st, 1960. The actual height was 83,523 when Andreev jumped. When Andreev and Dolgov were at 71,800 feet on the way up, Andreev wrote later in his book *The Sky—Around Me,* "We are at 71,800 feet. This altitude was reached for the first time in the world by Soviet aeronauts P. F. Fedoseenko, A. B. Vasenko, and I. D. Usyskin on the stratosphere balloon *Ossoaviakhim-1,* on January 30, 1934." (Translated from the Russian and shared by Jim Winker, Balloon Consultant)

Dolgov reached 93,970 feet, however, the helmet of Dolgov's pressure suit hit part of the gondola as he exited, and the suit depressurized, killing him. He was posthumously awarded a Hero of the Soviet Union. His height still did not exceed the record set by Joe Kittinger.

After Nick Piantanida's second failed attempt to pass both Andreev's free-fall record and Joe Kittinger's height record , F. C. Christopherson of the Argus Leader, in an editorial, "Of a Daring Breed to Whom We Owe Much," reflected, "Why He Did it. Many speculated in advance about Piantanida's venture. Some wondered just why he wanted to do it—a flight that obviously involved extreme risks. But his motivation is one we should understand and appreciate. He belongs to a grand company of adventuresome men whose eager desire to explore the unknown had been the force behind much of the world's progress." [11]

Nick and Janice discuss the glory and the danger before his
final jump from the stratosphere. – *Photo Courtesy of Joel Strasser*

"Nick vowed that on the next attempt he would go all the way to
140,000 feet, which must have been news to Jim Winker and the Raven
engineers who had already calculated the ceiling for the balloon they
were planning to build for the follow-up, and that was nowhere near
140,000 feet." [12]

Soon *Strato-Jump III* called *JADODIDE*, after Nick's wife and 3
daughters, (Janice, Donna, Diane, Debbie) was in the works. The launch
and planned jump took place at the Joe Foss field in Sioux Falls on May
1, 1966. The flight upward went well, until the Raven ground crew and
Ed Yost heard a "whoosh." Then a cryptic cry, "Emergen" and that was
it. Ed Yost, barked "Emergency, cut him off." The gondola hit the ground
in a wide open field 20 miles east of Worthington, Minnesota. Nick was
unconscious. My schoolmate at Augustana College, Kent Morstad, went
with the ambulance to drive Nick to the tiny Worthington hospital. From
there on May 2, Nick Piantanida was transferred to Hennepin County
General Hospital, where he was placed in a hyperbaric chamber. A

hyperbaric chamber is a pressure chamber where oxygen can be administered at a higher level than atmospheric pressure.

On August 30, 1966, Nick Piantanida died. The Argus reported "Nicholas Piantanida, who had clung to life for four months after an unsuccessful attempt to break the world's free-fall record, died Monday night. He was 33. Piantanida died at the Veteran's Administration Hospital in Philadelphia. He never regained consciousness." The Argus would write, "Nick Piantanida's daring adventure terminated unhappily. But it was a risk of which he was well aware and one which he was ready to assume in his effort to chart a new path in an important phase of scientific understanding." [13]

Ed Yost, project director for Raven Industries, said that there was no evidence of equipment failure and that it was possibly pilot error. Ed and I talked about many things in his Skypower shop at Tea, over the years, but Ed did not talk about Nick Piantanida. For all involved it was a sad, and no doubt painful event.

Author examines Nick Piantinida's parachutes. — *Courtesy of The Center for Western Studies at Augustana College, Sioux Falls*

In a July 5th, 2012 interview by myself with Raven Engineer, Jim Winker, Jim said, "Nick and his team should have had an extensive list of what could conceivably go wrong and both prepare for them and practice them. If he (Nick) had such a list, it was pitifully short." 14

The story of Nick Piantanida will be debated for some time to come. Was Nick a hero? Was he a visionary? Was he glory hungry with ego-driven ambition, leaving a wife as a widow and three children fatherless? Craig Ryan titles his book appropriately, on whatever fate decides – *The Magnificent Failure*.

A post script was added to this story, as history unfolded for a much better prepared Felix Baumgartner, on October 14, 2012. Felix Baumgartner flew to an amazing 128,100 feet, looked down and leaped. During his descent through the stratosphere, Baumgartner went into a dangerous out-of-control spin for about 40 seconds, experiencing nearly 2.5 times the force of gravity before stabilizing himself. According to reports in the public media, Felix achieved a free fall at 833.9 mph. He fell for 34 seconds before going supersonic. The actual jump to landing in the New Mexico desert took 9 minutes and 9 seconds. His jump was a *Magnificent Success!!*

Chapter Thirteen

Raven Rocks!

"Balloonists generally agree it was that 1960 flight that signaled the beginning of world wide hot air balloons."
– Raven's *"Celebration of 50 years of Innovation"*

It all began in 1956, when four pioneering entrepreneurs imagined a bold idea, born of individual inspiration and cooperative endeavor. Joseph Kaliszewski, James (J.R.) Smith, Paul (Ed) Yost and Duwayne Thon, employed at General Mills Inc. Aeronautical Research Laboratory in Minneapolis, Minnesota, left their employment to create a new company – **Raven Industries, Inc.**, in Sioux Falls, South Dakota.

In the beginning, the four innovators lacked capital. Ed Yost, working with the C. J. Hoigaard Company building nylon envelopes used for inflatable systems, and being tested at General Mills, approached General Manager Edwyn Owens and Owner

Kaliszewski, Smith, Yost, Thon
– Reprinted with permission from Raven

Cyrus J. Hoigaard about a new direction. Ed Yost would not only pique their interest but also gain their support. Since plastics were competing with the Hoigaard Company's product of canvas products, Owen and Hoigaard seized this new opportunity to gain expertise in the plastic industry. The added benefit included a balloon manufacturing and flight facility.

"What happened next, is part of South Dakota business history. A series of meetings took place with the four former General Mills employees, Hoigard, Owen, and H. P. Skoglund, brother-in-law to Hoigaard and President of North American Life and Casualty Insurance Company in Minneapolis. After much discussion and negotiation, an agreement was reached. On February 11, 1956 the Articles of Incorporation were filed with the Secretary of State. The group officially formed **Raven Industries** Incorporation." [1]

After an extensive search, Sioux Falls, South Dakota, was selected as an attractive site for this fledgling industry. The Joe Foss field had been used earlier for balloon launches. The winds were favorable in South Dakota and the topography suited balloon flights. Freight connections were available. No corporate income taxes made the Sioux Falls location perfect.

Along with money from Hoigaard, each of the four "pioneers" agreed to invest $5000.00 of personal money to get the company started. Joseph Kaliszewski and wife Beverly were the first to move to a rental home in Hartford, since rentals were not readily available in Sioux Falls. The others eventually followed.

Army barracks served as the first Raven Industries factory. Soon the first employee, Jerry Green, was joined by six engineers and technicians who began fabrication of plastic film products. Early research and production of scientific balloons paved the way for a young Raven Industries to start up and eventually thrive in a developing space age environment.

Ed Yost, having received a $47,000 grant from the Office of Naval Research, was able to design, build and launch the newly developed hot air balloon.

Before World War II, Jean Piccard flew the manned *Pleiades* from Soldier's Field, Rochester, Minnesota. The date was July 18, 1937. The *Pleiades*, using 92 latex balloons, lifted a gondola made of Alclad sheets. To descend, Piccard cut loose or shot balloons, until he safely landed a hundred miles from Rochester. After World War II, Jean Piccard and Otto

Winzen developed a balloon system called *Helios*, which was fashioned after the *Pleiades,* but was never actually launched. General Mills and the University of Minnesota soon became the focal point for all U.S. Balloon research and activity. The cold war was on, and the military spurred new interests in balloons. General Mills and the University of Minnesota benefited greatly from a New York team of meteorologists and physicists. "The NYU group had employed a specialist to fashion their envelopes from, what one participant fondly characterized as 'carrot bag grade polyethylene.' Having discovered the extent of the General Mills operation, however, it made sense for the NYU group to abandon their own balloon-making program and purchase their aerostats in Minneapolis.

Two original key figures, Piccard and Winzen, had left General Mills by 1949. Piccard had been drawn away from the balloon business by the press of other scientific interests. Winzen left the firm under a cloud in 1949 to found Winzen Research, Inc. With the help of his wife, Vera, who played a key role as vice-president and chief of production at Winzen, the firm quickly grew to become a leader in the field. Winzen would compete with Raven until 1994 when Raven purchased the Winzen assets.

"Charles B. Moore Jr., a member of the original NYU balloon group, replaced Winzen as head of the General Mills aeronautical laboratories. Other NYU alumnae, including J. R. Smith and H. A. Smith, the group's original polyethylene balloon builder, quickly followed Moore onto the General Mills payroll. J.R. Smith would be elevated to leadership of this group following Moore's departure for Arthur D. Little in 1953." [2]

(Actually, H. A. Smith's contribution was minimal and unrelated to any scientific group. H. A. Smith did manufacture a few crude polyethylene balloons for the NYU meteorological project.)

One of General Mills notable achievements, on May 17, 1954, was its *Operation Skyhook* balloon which reached a record altitude of 116,700 feet, which is more than 22 miles above the earth's surface. This balloon was the largest ever build up to that time – at 282 feet long when deflated, and 200 feet in diameter when inflated. Sent up to study cosmic rays, this huge balloon could be seen at distances of up to 90 miles.

"From the efforts of these and other pioneers, the plastic balloons that Jean Piccard conceived in 1935, began to play a major role in both science and national defense. To the general public, science seemed to be the prime beneficiary of the new technology. It is certainly true that the advent of a lightweight, reasonably low-cost of lifting instrument payloads to altitudes in excess of 100,000 feet, led to a renaissance of scientific ballooning." [3]

An October 2nd, 1960 article in the Editorial-Business section of the Minneapolis Sunday Tribune with the headline "Upper Midwest Is Balloon Capital of the Nation," highlighted the four major players in the balloon business. "The four balloon making firms – General Mills of Minneapolis, Winzen Research, Inc., of Bloomington, Minnesota, G. T. Schjeldahl Co. of Northfield, Minnesota, and **Raven Industries** of Sioux Falls, South Dakota – have easily made the area the high-altitude balloon capital of the nation." [4]

In 1955, Ed Yost landed the Office of Naval Research (ONR) contract for inventing and designing a manned balloon by tethering a plastic balloon, heating the envelope cavity using plumber's pots, burning kerosene, and ascending. The photos of the ascent convinced the Navy. The Navy had been engaged in exploring new balloon technologies for defense and scientific purposes. Gas balloons were expensive and limited in the amount of weight they could carry. The Navy was interested in an inexpensive, lighter-than-air, aerostat that could manage designated loads.

In 1956, Jim Winker was hired by co-founders Joseph Kaliszewski and J. R. Smith to analyze the data of a scientific balloon demonstration flight for another contract with the Office of Naval Research. "'The unusual flight with the payload carried on top of the balloon was completely successful,' reported Winker." [5]

Winker worked for Raven for 35 years and presently retains treasured, extensive archival Raven Industries historical data. At this writing, Jim Winker is writing a book on various technologies he was directly involved in while an employee. Jim Winker gave this account of the early years at Raven: "Back in the early years we were primarily a research and development company, and gradually became more of a production company. Enough customers transferred their balloon work from General Mills to provide most of our sales for the first five years." [6]

Eventually, Raven Industries diversified into a great variety of products such as plastics, electronics and parachutes–however this narrative primarily highlights ballooning discoveries, products and aerostatic achievements in South Dakota.

"In September, 1959, Raven completed the manufacture of a six million cubic foot scientific balloon, that when flown, established a world's altitude record of 150,000 feet. The balloon stayed aloft for 11.5 hours." [7]

In 1960, Ed Yost and the Raven team, experimented with a prototype craft called the *Vulcoon,* (an adaption from Vulcan, god of fire, and balloon). The *Vulcoon* was powered by a propane burner.

Then on a sunny Saturday morning, October 22, 1960, near an abandoned WW II bomber training base at Bruning, Nebraska, Ed Yost flew the world's first modern hot air balloon–*ONR – Mark I.* With Ed were Jim Winker, Gino Manchuso, Darrel Rupp and two observers from the Office of Naval Research.

Ed Yost prepares for his first maiden flight.
– Photo courtesy of Jim Winker

Ed flies the first modern hot air balloon. — *Photo courtesy of Jim Winker*

Mark I, constructed of nylon with a mylar film laminated to the interior, was 40 feet in diameter, containing 30,000 cubic feet of air. There was no basket gondola, but a boatswains chair to sit on. On this maiden flight, the Raven Team walked the fledgling aerostat downwind and soon the balloon and pilot Ed became airborne. The balloon flew to a height of 500 feet, and after 25 minutes in the air, landed three miles away. Was there a post flight celebration one might ask? "Not immediately, no," said Jim Winker. "I think everybody was relieved and pleased that it had flown. Everybody realized too, that it wasn't the performance that we needed to get out of it before the project was done, but it was a major milestone. The moments of elation were brief, people just went about their business cleaning it up, packing it up and getting ready to go back home." [8]

The burner is the key to a hot air balloon. The burner is the engine as it burns propane in this new modern craft. Kerosene was too temperamental and had the tendency to breakdown and leave a coating on the inside of the burner, gumming it up. With propane—after the pilot light is lit, and the metal coils begin to heat up—the propane heated in the coils becomes gas. When propane is burned as gas, it creates a powerful flame producing sufficient British Thermal Units with amazing lifting power.

Ed Yost flies 2nd flight out of the Stratobowl..
— Photo courtesy of Jim Winker

Three weeks later, on November 11th, the team was ready for the second manned free flight, featuring a new, improved balloon – the *Mark II*. For this flight, the Stratobowl in South Dakota was selected. November 11th was also at the same day that Joe Kittinger and others were celebrating the 25th anniversary of the flight of *Explorer II* up on the north rim. On November 12, 1960, weather conditions were ideal. This time, using a larger fan, and a weed burner to preheat the air, the inflation took only six minutes. The balloon took off, again with Ed Yost as pilot, and climbed at 400 feet per minute to 7,000 feet. At this altitude, Yost closed the fuel valve a few turns, reducing the burner output. This resulted in a leveling off, which was maintained for about 20 minutes with minor adjustments. Then the burn rate was increased again and the balloon soared to higher altitudes, climbing at 300 feet per minute and reaching 9,000 feet. When Yost began to close the valve this time, he closed it a bit too much and a rapid descent (600 feet per minute) was initiated. He applied full heat to recover and the balloon, after having lost 1000 feet, now began to climb rapidly. These oscillations continued for 15 minutes while Yost gathered experience in controlling them. Finally Yost regained level flight. Then he reduced the heat slightly to permit a gradual rate of descent. (For more specific details of these flights and including the twenty-six ascensions made during the period beginning April 3, 1961 and ending October 19, 1961 consult the *Final Report: One Man Hot-Air Balloon System Development for Office of Naval Research, Report 1863 on November 1, 1963 prepared by P. E. Yost, Program Manager, Raven Industries, Inc. Sioux Falls, South Dakota*). "'I had a parachute on. I didn't put this in the final report, but that thing was going up and down for a while, so wild, I almost bailed out. Then I said to myself, you dumb fool, it hasn't hurt you yet! Sit back and enjoy it. I finally got her halfway leveled out, and made a long gradual descent.' (Yost)....after one hour and fifty minutes of flight and covering a distance of 39 miles, Yost accomplished a normal landing. 'This flight was a beauty,' Yost said.

This second flight and further test flights over the next years led to a modification of the fuel valve to provide an immediate on-off response. The vaporization jacket surrounding the flame was replaced by seven coils of stainless steel tubing. A skirt below the mouth and a maneuvering vent were also invented and added. With these improvements, all the essential ingredients of the modern hot air balloon were in place." [9]

Ed Yost had began experimenting with hot air balloons in the fall of 1954 on his own. First, he constructed a 10 foot envelope. Then Ed lit a plumbers fire pot under it. Then he went on to construct a 30 foot envelope and used three pots. Next, he used 5 fire pots under a 39 foot, 27,500 cubic foot balloon. When he inflated the balloon using a plywood seat and a foot tire pump to pump air pressure into the white leaded gas tank, at Huron, South Dakota, it actually flew. That tethered effort paved the way for the $47,000 grant from the Office of Naval Research.

With *Mark I* and *Mark II* flights, Raven was on the way to an unknown but exciting and profitable future. In 1961 Raven sold its first sport balloon to Dr. William McGath, and the popularity of ballooning blossomed. Don Piccard, the son of the renown balloonists Jean and Jeanette Piccard, was hired as Marketing Manager by Raven to promote and sell the new product. Don Piccard came from a family of balloonists – his uncle, Auguste, who invented the bathyscaphe (actually an underwater balloon sphere) and his father Jean, who was a twin to Auguste. Don Piccard's father, Jean flew a gas balloon to over 57,000 feet in 1934, the year before *Explorer II*. Don's mother, Jeanette, an Episcopalian priest, became the first woman to fly a balloon in space. The Piccards, Stevens, Anderson, Ed Yost and others produced many **firsts** in U. S. scientific aerial history.

Chuck Yeager broke the sound barrier when he became the **first** person to fly an airplane faster than the speed of sound. Jackie Robinson broke the color barrier in major league baseball when he became the **first** black player for the Brooklyn Dodgers. Roger Bannister became the **first** athlete to break the four-minute-mile barrier when he ran the distance in 3:49.4 at Oxford. Colonel Joe Kittinger became the **first** person to ascent into the stratosphere and parachute down faster and further than any other human being, and land safely. On October 14, 2012, Austrian Alex Baumgartner jumped from 128,100 feet in a parachute, and became the **first** to break the sound barrier with his hurtling body and also land safely. Paul "Ed" Yost became the **first** person to invent the modern hot air balloon while serving as the Operations Manager of Raven Industries in Sioux Falls, South Dakota. Acknowledging the Raven Team effort, Ed has been dubbed "the *Father of the Modern Hot Air Balloon."*

"Raven Industries touting 'the Most Beautiful Balloons in the World' enjoyed snowballing market share throughout the 1960's and 1970's and beyond, and remained a major force competing with other balloon manufacturers until 2007 when the company subsidiary, Aerostar, ceased accepting new balloon orders." [10]

Ed checks propane tank prior to first hot air balloon flight across the English Channel with Don Piccard. — *Photo courtesy of Jim Winker*

In 1963, the two ballooning legends, Ed Yost and Don Piccard, became the first to cross the English Channel in a modern hot air balloon launched from England, and landed in France. The *"Channel Champ"* as the areostat was named, is presently on display at the National Balloon Museum in Indianola, Iowa. The flight proved the viability of this new generation of hot air balloons.

In the 1960's, Raven constructed several balloons that flew in Hollywood movies. *Five Weeks In A Balloon* by 20th Century Fox was one of them. Another was *Skidoo,* produced by Paramount Pictures, which featured a balloon that was designed to appear "prison made" from old feed sacks and bed sheets. Used in an Alcatraz escape scene, it was flown by Raven Vice President, Ed Yost, on location off a barge adjacent to Alcatraz. Jackie Gleason and Carol Channing starred in it. *The Great Race,* with Warner Brothers, starred Tony Curtis and Natalie Wood was another Raven-made movie balloon. Ed told this writer that he "flew *The Great Race* balloon, crouching down unseen in the basket."

Raven blimp products at Macy's parade – *Reprinted with permission from Raven*

"In 1967, Raven formed a new company with Bohemia Lumber Company of Oregon. The new company called 'Balloon Trans-Air Inc,' manufactured and sold heavy load, short haul balloon transport systems for logging, construction and maritime ship-to-shore uploading operations." [11]

"1968 was the year a Raven balloon set a world record for duration by remaining aloft 441 days. Balloons of that type would circle the earth dozens of times, providing meteorological information not available from any other source. One of Raven's new products, a super-pressurized Mylar sphere, circled the earth four times near the earth's equator." [12]

In the second decade – 1966-1975, one newsworthy project that caught the public's eye took place when Raven's Applied Technology Division designed and constructed Mylar balloons to be used by the magazine mogul Malcolm Forbes, as he planned a crossing of the USA and Atlantic Ocean. Ed Yost served as operations manager for the land crossing. The Atlantic Ocean crossing attempt failed.

As the popularity of sport ballooning caught on, balloon races became increasingly popular, throughout the United States and the world. Raven not only manufactured aerostats for the general public but also was actively engaged in racing, having captured six of the top eight places in the 1972 National Balloon Races. In 1980, Raven balloons won both first and second places in a field of 32 contenders in the first International Balloon Championship. Raven also supplied the official balloon for the 1980 Winter Olympics. "Another pleasant surprise was the success of the new advertising blimp. Called the 'AdverBlimp,' sales from these large unmanned helium balloons grew steadily, and soon the production line was expanded to include a smaller version for use in retail stores." [13]

By February 1, 1986, it was time to specialize. Whereas Raven produced many products other than balloons, Aerostar International was established as a wholly-owned subsidiary of Raven to focus on the manufacturing and sale of sport balloons. Aerostar's mission was to continue pioneering work Raven had done with hot air balloons. "Mark West became President of Aerostar, following Mike Nystrom, and the chief engineer for hot air balloons. Mark West helped advance the science with new designs, especially in fabric construction." [14]

"In 1987, Raven was named 'One of the best small companies in America' by *Forbes* magazine while achieving both record sales and profits. This naming happened a second time and then in 1992 for the third time." [15] Later Raven was named again, "One of the Best..." for 6 straight years from 2005-2011.

Over the years, Raven's production and sales included plastics, clothing, parachutes, and from its newly formed Flow Controls Division, sprayer controls, chemical injection systems, valves, pumps and other precision mechanisms for agriculture. Raven prospered over the years because of excellent leadership through its directors, officers and division managers. Orv Oliver, Russ Pohl, Ray Ramstad and Jim Winker are but a few of the leaders and employees that enabled Raven to succeed.

"Raven then became involved with project *Earthwinds*. Its mandate was to design and manufacture a helium-filled balloon for an around-the-world flight. This embodied an entirely new concept of how balloons could be used for near space research. *Earthwind's* design involved using two balloons – one helium-filled balloon made of Raven Astrofilm, connected to a second anchor balloon which served as ballast. 'But the project ultimately never achieved its full potential,' remembered Raven's Jim Winker." [16]

In 1994, Winzen, a chief competitor in the field of high-altitude balloons, was purchased by Raven Industries. Raven became the primary supplier for high-altitude balloon research.

Jim Winker retired in 1991. He is presently one of the most knowledgeable experts not only regarding the history of Raven Industries but as a technical ballooning consultant.

"In 2005, Aerostar International successfully launched and flew the second airship in history to achieve powered flight in the stratosphere. This was launched in cooperation with the Southwest Research Institute and the US Air Force Research Lab. The project, titled, *HiSentinel,* launched from Roswell, New Mexico, focused on developing a small inexpensive tactical communications and Intelligence Surveillance and Reconnaissance applications. The first one, which flew in 1970, was also a Raven product." 17

In 2007 Aerostar ceased hot air balloon operations. However, new products in other fields enabled the company to thrive to the present day.

Raven's balloon legacy in South Dakota is unparalleled. The dream and vision of 1956 by four imaginative pioneers, has surpassed even the most adventuresome thinkers.

Chapter Fourteen

Skypower

*"All other sports cannot compare! Ballooning has every-
thing – the constant challenges; the unusual excitement it
creates, even to the ground observer; the aesthetics and,
of course, it is also a team sport. What more can anyone
ask for?"* — Malcolm S. Forbes

Little did the residents of a sleepy little town near Sioux Falls, South
Dakota, realize that a legendary balloonist lived and worked among
them. Quietly and without fanfare he made history.

It was in 1968 that Ed Yost left Raven Industries in Sioux Falls, the
company he co-founded, to form his own company with Dick Keuser
– Dakota Industries Inc., in Tea, South Dakota. The Vietnam War was
raging. The U.S. military needed parachutes. Ed's Dakota Industries
received contracts and produced them by the thousands. Soon however,
that gray corrugated turtle-shell-like quonset structure, modest in appearance, would house the designing and constructing of history-making
gas balloons. Dakota Industries eventually became 2 separate entities
– Universal Systems and Skypower.

Ten years later in 1978, I walked into that Quonset building, answering
an ad for a draftsman. "Howdy," a spotted brown and white Springer
Spaniel dog greeted me. Ed Yost sat on a stool barely looking at me.
"What's your qualifications," he asked rather gruffly. "I worked for
Boeing," I answered. "I was assigned to work on the Bomarc Guilded

Missile project as a draftsman," I continued. I stood silently in awe and just a bit afraid, for I already knew of his stature and famed ballooning adventures. "You're hired." he said. "What?" I thought, "no resume, no letter of recommendation, no negotiations, no drug testing – just 'you're hired?'" "Here," he said. He handed me a four inch maple toggle. "Draw this." I went home to my drafting board and I drew it. I returned, Ed muttered, and that was the beginning of a 30 year long treasured relationship. I learned later that Ed himself was a 1940 graduate of the Boeing School of Aeronautics in Oakland, California. Were we "Boeing Brothers?" I immediately went to work drawing blueprints for the gas balloon series GBN-41-1000 for FAA Type Certification. Later, when finished, the FAA representative told Ed and then Ed smiled and told me, "those blueprints were the best they had seen." Ed was building helium gas racing balloons, rated at 1,000 Cubic meters, designed to carry old style wicker baskets made from natural Rattan Reed called Koo Boo. Ed wanted manual drawn (not CAD) blueprints, drawn in classic style like the prints of the past. Those wicker baskets would mimic the style of gondolas used in the early days of the Gordon Bennett World races and even earlier.

I would confer with Ed, Suzie, Jerry Melsha and Dick Keuser. I would collect measurements and data, and then race home to my drafting studio in nearby Harrisburg. In my spare time and in off

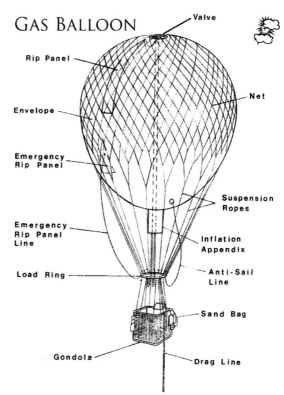

A typical Gas Balloon manufactured at Skypower
– Sketch by A. Fadness

GAS BALLOON

Valve
Rip Panel
Net
Envelope
Emergency Rip Panel
Suspension Ropes
Emergency Rip Panel Line
Inflation Appendix
Load Ring
Anti-Sail Line
Sand Bag
Gondola
Drag Line

hours I drew furiously and delightfully. I worked in those days full time at Shalom Lutheran Church. As a pastor I was immersed in theology, sermons and the business of the church, so this drafting project became my free time hobby. I loved the challenge and exposure to the world of ballooning. Little did I know what amazing ballooning experiences I would see in the future. It always would tie in with Master Balloonist E. Paul Yost.

I had worked as a draftsman in the early boom days at Boeing's – Plant II – when the 707, the KC135, (still flying today), and the 2 hundred mile range Bomarc Defense Guided Missile were designed and put into production. The Cold War was hot. Military projects were receiving the highest priority.

Later, during college I drafted blueprints for Spitznagel Architects, Howard Parezo Architects, and Ken Bastian Engineering in Sioux Falls. While a student at seminary I drafted for Gausman and Moore Consulting Engineers in St. Paul, Minnesota and Boonestro, Rosene and Associates Civil Engineering also in St. Paul. In the parish I drafted cemetery plots, home blueprints, and inventions for inventors. My love for drawing, drafting, graphic arts and calligraphy blossomed as I worked at Skypower.

Fadness securing data for Wicker basket blueprint

Working with Ed Yost, his wife Suzie and the crew at Skypower, provided not only a window into the art and science of ballooning, but also an interesting study in psychology. Ed was a no-nonsense, brilliant leader. He was direct, cryptic and impatient with trivia. Once he barked at me with "don't sweat the small stuff." Suzie, not only his beloved wife but also loyal companion and team partner, was the communicator, liaison, mediator, and translator. Suzie brought grace and elegance to the Skypower shop. In Susanne Robinson Yost, described as a "tall, beautiful and brilliant woman who drove sports cars," Ed Yost met his match. Suzie had passed the bar exam without ever attending law school, and later became a part time judge in California. Suzie met Ed when crewing for one of Ed's balloon flights. She and Ed were destined to be a team. At Skypower, Suzie was called "the Spider-Woman" by friend Colonel Joe Kittinger. She specialized in making the nets for the thousand cubic meter gas balloons. My role was to draft the designs and details of the nets which connected envelope to load ring, load ring to wicker basket. Our symbolic engineer was Clyde Pritchard. Dick Keuser and Jerry Melsha directed most of the construction of these special aerostats. Later, a second gas balloon of 500 cubic meters was designed, drafted and produced. It was Skypower's Model GB N-32-500.

Ed Yost prepares a launch at the Thunderbird Gas Balloon Rally in Phoenix, Arizona, in 1984. – *Photo courtesy Ron Behrman Photography*

All Yost-built gas balloons in flight. — *Photo courtesy Ron Behrman Photography*

During the Skypower years, 1968-1987, Ed Yost contined to make history in the balloon world. It was in 1969 that Ed Yost, Don Kersten and Peter Pellegrino organized The Balloon Federation of America. Originally the BFA was organized as a gas balloon organization. Don, Ed and Peter proposed to the National Aeronautic Association, a new constitution, which changed the BFA into a hot air ballooning federation. However Ed had a "checkered" past when it comes to the BFA. It wasn't long before the officers of the BFA were making rules for whatever reasons and possibly for their own benefits. Ed Yost found himself thrown out of his own organization for flying in an unsanctioned balloon event in Canada. So what did the erasible Ed Yost do? He enrolled his dog in the BFA, claiming that man's best friend was the "guard at his Skypower balloon factory." In the dog's name, Yost would send in letters critical of the way the BFA was run. Yost tells how the editor of the BFA magazine was on to him, but ran the letters anyway. At one point, Yost's dog ran for the BFA board and garnered more votes than some of the other candidates who failed to win. Eventually, all was forgiven and Ed was sent a plaque declaring him a Life Member.

Ed Yost preparing his next project. — *Photo credit Ron Behrman Photography*

On August 17-23, 1970, Ed Yost organized the first U.S. National Hot Air Championship at Indianola, Iowa, 130 miles from his hometown of Bristow, Iowa. The site was wisely chosen. In that idyllic, flat, farmland, and with great citizen support, a balloon festival was launched which has flourished to the present day. The National Balloon Museum was established at Indianola, Iowa and has become an attractive tourist destination along with the festival.

In the early 1970's Ed Yost organized the first World Hot Air Ballooning Championship in Albuquerque, New Mexico. He wrote the rules and served as clerk of the course, which is an honored position now called the Balloon Meister. This World balloon ascension attracts hundreds of balloons and balloonist annually. It has become the premier balloon event in the world.

One day in 1973, the Skypower phone rang. It was Malcolm Forbes, the Editor-Chief and Publisher of *Forbes*, the nation's leading business magazine. Forbes was a collector of art, a motorcyclist and a licensed hot-air balloonist. He asked for Ed Yost. Ed answered. Without unnecessary small talk Forbes said, "I want you to build me a balloon to fly across the United States. When are you coming to New York?" Ed replied, "Never, if I can help it." To an inquiry about when Ed would be visiting other eastern cities, Forbes got the same answer. "Never!"

Prior to this conversation, Malcolm Forbes, the millionaire magazine magnate, had gotten into ballooning almost by accident. "He was nestled back in the cushions of his huge maroon Mercedes-Benz limousine, on the way from rural New Jersey into Manhattan, when he read of the nearby trial balloon flights. *Sports Illustrated* reported what happened next. 'It sounded like a cool idea,' Forbes said, 'so I talked my chauffer into going up with me. The next thing we knew we were both signed up for lessons.'" [1]

Malcolm Forbes was not one to say no to Ed Yost's rebuff. "About two weeks later, Forbes called again. Yost told him he was not interested. Not long after when Ed was working on a project in Louisville, Kentucky, he heard a knock on his hotel door. It was 10 a.m. on a Saturday morning. There stood Malcolm Forbes. When Ed asked Forbes to help him unload some boxes, Forbes protested. He lamented that he was too old. They compared ages. Ed was a month older. Forbes went to work.

Ed Yost flies Malcom Forbes across the continental United States. – *Photo Courtesy of Joel Strasser*

Malcolm Forbes wanted to fly from San Diego in July, in 14-15 hours a day, for a transcontinental flight. Ed, not one to comply to other's opinions especially when he was the expert, suggested flying from Tillamook, Oregon, (where Ed had worked on projects in a WWII blimp hanger) starting in October. Forbes was concerned that his Convair plane, which would be used as a chase and support vehicle, needed to be able to fly to any Oregon launch site. Nearby Coos Bay would be selected.

Ed built the balloon and finished it in August. It was designed to be taken apart to fit in Forbe's twin-engine Convair airliner. The balloon was dismantled and flown in the Convair in order to fly to Forbe's Chateau de Ballory in Normandy, France.

In France, one day, Forbes flew the new balloon along with another pilot flying a French balloon. Forbes wanted it to be a friendly competition, but he wanted to win. He also wanted it to look like he was flying the balloon even through Ed Yost was actually piloting it. When the French balloon landed, Forbes and Yost landed about a hundred yards beyond just enough to "win" the informal competition.

Then it was time for Forbes and Yost to go back to the United States, for the planned cross country flight. Forbes had a huge mobile home that Ed called the Taj Mahal. Forbes went along with the "crack" and painted "Taj Mahal" on the side of the vehicle.

"Yost, and about half a dozen of his most experienced people, flew Forbes for the first part of the adventure until Forbes became experienced with the balloon. In Nebraska, Forbes had to inflate the balloon and take off from a field strewn with hay bales in high wind conditions. A local woman had parked her used but beautiful and prized Cadillac along a road downwind of the balloon to watch the action. Bouncing off of hay bales, the basket hit and caved in the side of the Cadillac before Forbes got airborne. The woman was distraught, to say the least, and was crying inconsolably. Forbes' crew radioed up to him what had happened. Forbes told the crew to buy the woman a new Cadillac!

At the end of the flight, Forbes had to land the balloon in the Chesapeake Bay in high wind conditions. Ed said that Forbes thought he was going to drown, but when the balloon hit the water and Forbes got out of the basket, he found that he was in the shallows and that the water was only up to his waist!" [2]

The stories, like the Forbes-Yost adventures and others, are many – far too many to capture in this aerial monograph. Nonetheless, it was out of a tiny company called *Skypower,* along with *Universal Systems* in Tea, South Dakota, that a ballooning legend and his cohorts made magical history.

Chapter Fifteen

The Gordon Bennett Races Connection

"Then you have discovered the means of guiding a balloon?" "Not by any means, that is a Utopian idea." "Then, you will go ——Wither-soever Providence wills......."
— Jules Verne, *Five Weeks in a Balloon*, 1862

In the late 70's and early 80's, the Skypower crew at Tea, South Dakota, rolled up their sleeves and went into action. Ed Yost concocted a new mission. Skypower would design, produce and market netted classic style 1,000 cubic meter balloons for gas balloonist enthusiasts and for the revived International Gordon Bennett Balloon Races. Key players on the Skypower stage working to produce this new aerostat would be CEO Ed Yost; Susie Yost the net maker (Joe Kittinger called her the Spider Woman); Clyde Pritchett, temporary engineer; Dick Keuser and Jerry Melsha supervising the balloon construction; myself, Arley Fadness, the drafter; and several other factory workers. I would be drafting manually, old style prints, reflecting the early balloons, for Type Certification for the FAA. This was the beginning of a long and treasured relationship with Ed and Suzanne Yost and myself.

The envelope, netting, load ring, valve, rip panel and wicker basket would resemble as closely as possible the aerostats that flew from before 1906 and on. Some have asked, why the old style wicker baskets when other more modern styles were available?

"Wicker construction has an advantage over metal skeletons and hard fiberglass shells in the absorption of kinetic energy of impacts – however wicker is also favored for it's nostalgic artistic appearance." 1

Skypower would manufacture at least 17 aerostats for sport and specifically for the historic Gordon Bennett Balloon Races. The cost for these helium lifting balloons would be around $30,000.00 with a $1,000 cost per training flight. This would obviously be a sport for the wealthy.

When one takes a look at a bit of ballooning history, one begins to understand the excitement that went on in that corrugated quonset factory on the edge of Tea, a sleepy little town in South Dakota.

"By the early 1900s ballooning was no longer the exclusive province of scientist, showmen and the military. Well-to-do sportsmen and women on both sides of the Atlantic had embraced it as an entertaining diversion; weekend jaunts and races – won by the balloon that traveled furthest or that landed closest to an appointed spot–became fashionable pastimes at resorts and suburban parks.

Ballooning for sport got its start in France, where the wonders of lighter-than-air flight had first been demonstrated in the late 18th Century. But society balloonists enjoyed their finest hour in Edwardian England. There, from the well-tended grounds of exclusive polo clubs such as Ranelagh and Hurlingham, both near London, elegant aeronauts enlivened the summer season with free-flight competitions and pleasure voyages, frequently provisioned with fine food and chilled champagne.

Bearing well-dressed merrymakers serenely through the air, the early sport balloons were symbolic of that gilded, carefree era before Europe's old order was shattered by a devastating world war. Yet their popularity prestaged a later day when ballooning for fun would become a pastime for thousands of people of every social class." 2

The impetus for this new zeal and enthusiasm for sport ballooning events and races came about in large part from an eccentric New York newspaper publisher — James Gordon Bennett. Bennett was a flamboyant, stylish personality who established prizes for airplane, automobile and yacht racing. They called him "Commodore" because of his love and involvement in Yacht races.

The "Mount Everest" of all sport balloon racing capturing the imagination of wealthy Americans, as well as the European elite, was the newly established *Gordon Bennett Gas Balloon Races*. "Twenty-six

Gordon Bennett races were staged in six different nations between 1906 and 1938. The United States won ten of these contests. Belgium seven. Poland four. Germany and Switzerland two each, and France one. A grand total of 351 pilots and copilots, the world's foremost aeronauts, participated in the competition. One hundred thirty-nine men competed in more than one of the contests. Ernest Demuter held the grand record, however, flying in eighteen of the annual events." [3]

The Gordon Bennett Race was actually a distance contest. The balloon envelope size ranged from 22,000 to 80,000 cubic feet in capacity. The balloons, over the years, were inflated with coal gas, hydrogen or helium.

The very **first race,** September 30, 1906, was an international hit. Sixteen balloons from seven nations raced. One millon cubic feet of coal gas filled the sixteen balloons. Two balloons represented the United States – Alberto Santos-Dumont, who flew the only six horsepower machine (ever allowed) in the *Deux Ameriques,* and Frank Purdy Lahm. Lahm's last minute copilot selection was Henry Blanchard Hersey. "Lahm, Hersey, and the other contestants began their competition from the Tuileries Gardens in Paris. An enormous crowd of spectators had gathered in the garden and the neighboring Place de la Concorde hours before the scheduled 4:00 p.m. launch. Lahm recalled that Parisians crowded the nearby bridges over the Seine, hung out the windows, and darkened roofs.

Several bands entertained the crowd with the music of Offenbach. The festivities were enlivened by the release of hundreds of small colored balloons and of inflated figures constructed of goldbeater's skin. Immediately prior to launch, a large flock of pigeons was released. As Lahm recalled, "All of this, combined with the natural gaiety of a French holiday gathering, made a fascinating and long to be remembered scene." [4]

"Lahm and Hersey in this first race, crossed the English Channel, which Lahm said, was the 'most interesting part of our voyage.'" [5]

Lahm recalled, "To describe the beauty of a Channel crossing would require the pen of a master. With a full moon shinning overhead, an almost cloudless sky, the balmy air, and except for the gentle breaking of waves beneath us, not a sound to disturb the perfect calm, nothing could be more charming, nothing more delightful. With occasional reference to the compass and north star, we knew our direction was good, so had no uneasiness on that score. Sitting on the bottom of the car on the ballast bags, occasionally looking over to see if the guide-rope was clear of the water, if not, throwing out scoopful of sand to send us up a few feet, we quietly ate our long-postponed dinner of sandwiches, chicken, eggs, fruit, coffee and other good things which we laid in before starting. Once a little sailing vessel slipped under us and disappeared in the night. This was the only sign of life we saw in the Channel." [6]

Lahm and Hersey traveled 402.4 miles in 22 hours and won the first James Gordon Bennett Coupe Internationale de Aeronautes. The winner of the Gordon Bennett Race brought the trophy to the home country with the privilege of hosting the race the following year.

The **second race** on October 21, 1907, was launched from St. Louis with nine entries. There were three from Germany, two from France, one from England and three from the United States. 300,000 spectators gathered for the launches. Oscar Erbshoh, the German aeronaut entry, won achieving a distance of 872 miles which broke the American distance record which stood since 1859 when Professor John Wise flew from Forest Park, St. Louis to Henderson, New York. [7]

This second race was an enormous and complete success. The American aeronaut team, J. C. McCoy and Charles de Forest Chandler finished fourth. The other two American teams finished 5th and 8th.

The **third race** took place in Berlin, October 11, 1908. In this race, the Americans, Forbes and Post, lost much of their ballast in a takeoff accident, and the balloon went skyrocketing to 300 feet where it burst. While hurtling to earth at breakneck speed, Forbes cut ropes, allowing the bag to form an improvised parachute inside the netting. This broke the fall somewhat, but they crashed through the roof of a house, stopping suspended above the dinning table of a dignified German couple at lunch. [8]

During one year's celebration, a story has been told of the American who went ballooning with a foreign entrant. When the foreign aide was injured and unable to take part fully in the race, matters became complicated because neither man knew the other man's language. They had been up all night and had been using their drag rope to conserve ballast. When the rope became tangled in a tree, the foreigner dived into his luggage aboard the basket, muttering curses, and pulled out a stick of dynamite. The man jumped onto the webbing of the huge balloon as the American stood frozen with fear. What he feared most, was that disappointment had temporarily deranged his companion, for within a few feet of the basket was a bag filled with 80,000 cubic feet of highly inflammable hydrogen! Nevertheless, the foreigner hurled the dynamite downward with a practiced flip. It landed in the tree and did its job cleanly and neatly. The balloon was freed. [9]

Post and Forbes finished twenty-third in a field of twenty-three. Swiss aeronaut M. Schaeck traveling 808 miles in 73 hours and won the race and the honor for the next race to be in Zurich.

The United States won **race four** in October of 1909. Edgar Mix and his companion M. Roussel traveled 696.5 miles. Their flight encountered hazards. They became a target of hunters near Warsaw. They were stopped near Prague by a group of young men holding on to their dragline.

On September 17, 1910, **race five,** with only ten entries launched from St. Louis. Swiftly the balloonists flew into the danger zone of the Great Lakes. Seven of the ten crossed into Canada and temporarily disappeared. The German crew, Lieutenant Hans Gericke and Samuel F. Perkins, flying *The Dusseldorf,* remained in the air for 42 hours, 20 minutes, landing 17 miles north of Lake Kiskisink, Quebec. They were declared victors at 1,127.5 miles. But the *America II* crew of Hawley and Perkins who eventually won at 1,172.9 miles, were lost. Gericke and Perkins had no idea where they were. Once on the ground, they began to realize the seriousness of the situation. The underbrush was so dense that, without a hatchet or cutting tools, they were lucky to hike a mile a day. They crossed three small rivers and two lakes on makeshift rafts. Each night they camped on the highest point they could find, building a huge bonfire. One man slept while the other tended the fire. "We heard wild beasts all around us," Perkins recalled, "and thought we could distinguish the howling of wolves, but nothing attacked our camp." Finally, at noon on Saturday, October 22, they heard a gunshot. Perkins scrambled up the nearest hill and began to wave a red flag and fire his 22 pistol into the air. Theodore Baldwin, a local game warden, saw the signal and took the aeronauts to the nearest settlement in his canoe. [10]

By October, 24, Hawley and Perkins, aboard the *America II,* were still missing. Then a week after takeoff, the missing and exhausted aeronauts appeared. They had wandered for five days in the Canadian forests before being rescued by trappers. Keeping aloft for 46 hours and traveling 1,172 miles in a direct line—-they had established a new record.

The **sixth international race,** the smallest ever, flew out from Kansas City. The German Lt. Hans Gericke won the trophy and brought the next race to be staged at Stuttgart, Germany.

Race seven was a disaster for American contestants. One balloon, the *Kansas City II,* split during inflation. Another team was imprisoned in a Russian jail when they landed on Russian soil.

"But it was an all-American affair back in Tuileries Gardens in Paris in 1913, the same setting as the first Bennett classic. Ballooning by now was so popular that the race attracted more than a million spectators." [11]

The trophy for **Race eight**, was awarded to the *Goodyear* balloon piloted by Ralph Upson. (Cf. Ralph Upson's book *Captive and Free Balloons* published in 1926 for specific details).

H. E. Honeywell and his aide, Herman F. Long of Kansas City flew the *Uncle Sam* to the northeast, passing over Warsaw and then on over Russia, Honeywell noted, was a "desolate country compared to prosperous Germany, things had begun to look very dismal." The howling of wolves on the ground was a final touch. As Honeywell commented, "It was trouble from there on to the end." [12]

By 6:00 a.m. on the third day, the *Uncle Sam* "resembled an iceberg." They began dropping rapidly. Everything went overboard, but the aeronauts could not halt their descent. The flight ended in a grove of dead trees. "The bag was a total wreck." The Americans soon found themselves surrounded by "hundreds of peasants, some with guns and axes.....None could read their own language, much less talk ours. We drew a picture of a horse and wagon. They shook their heads. We then rubbed our stomachs, chewed our thumbs and pointed up and down the railroad track. One long haired fellow pointed west, holding up five fingers, by which we understood 'five verstes to town.'" [13] (Note: A verste is an obsolete Russian unit of length to about 1.067 kilometers or 0.6629 miles)

After two more hours spent floundering through deep snow, they arrived at a railway station, where they discovered that they had flown 1,054 miles to capture third place in the Gordon Bennett race. Their troubles were far from over however. After returning to the landing site with a handcar to retrieve the remains of the *Uncle Sam*, they boarded a train for Germany at 7:30 that evening. The aeronauts were stopped short of the border, dragged off the train early the next morning, and sent back to a provincial center under guard. They were searched and their belongings were seized. No contact with the U. S. consul in St. Petersburg was permitted. They paid for their own meals at the station restaurant and slept on a desk in the room where they were locked up at night.

Three days after their arrest, a wire came from St. Petersburg ordering their release. Reporting the experience to American newsman, Honeywell could only breathe a sign of relief. "Thank God we got out and are home once more." [14]

John Berry and his assistant, A. von Huffman, had a less adventurous and much shorter voyage aboard the *Million Population Club*. Berry was an especially colorful fellow. A native of Rochester, born in 1849, he made his first ascension at age fourteen. He moved to St. Louis where he became involved in an airship scheme. When Berry was shot by a coworker in an argument over this project, he sued and purchased his first balloon with the settlement. [15]

July 28, 1914 would end the International Gordon Bennett Balloon Race contests with the assassination of Archduke Franz Ferdinand of Austria by a Yugoslav Nationalist. World War I erupted. A much deadlier international contest than the Gordon Bennett Races ensued.

Even though The Gordon Bennett Races and other sport ballooning events were suspended, nonetheless, their history and experiences offered invaluable and much needed technological ballooning devices and experiences for use in the military.

Happily, the Gordon Bennett Race resumed October 23, 1920. At the **fourteenth** National Balloon Race, which always preceded the Gordon Bennett **race**, tragedy struck in 1923. The envelope of Navy balloon, A6698, flown by Lieutenant L. J. Roth, aided by Lieutenant T. B. Null, was found and picked up by a Lake Erie tug 25 miles south-southeast of Port Stanley, Ontario. After a full-scale search by two DH-4s, a Loening Air Yacht from Selfridge Field, and the flying boats, *Buckeye* and *Nina,* of Aeromarine Airways, the basket was found floating on the lake with Roth's body strapped in place. Lieutenant Null was never found. [16]

Three ballooning stars stand out in a number of Gordon Bennett balloon races. They were Roy Donaldson, Ralph Upson and Ward T. Van Orman.

Van Orman tells in his book, *The Wizard of the Winds,* about a har-rowing experience in the 1925 race. He and his aide, Woolam, inflated their Goodyear Balloon from Solbush Field and launched from Brussels,

Belgium. They flew over England and straight out into the Atlantic. There was no chance to land on terra firma. "Quite suddenly an idea kept coming to mind which, at first, seemed so ridiculous that I discarded it. It was this: 'If you don't want to be disqualified for descending into the sea, and have no land on which to alight, there is only one choice left, and that is to descend on a boat.' As wild as the idea sounded we were forced in our dilemma to consider it." [17]

Van Orman spied the *S.S. Vaterland,* a reparations ship of World War I. By extraordinary skill they landed on the 25 foot deck of this tiny German freighter. It had not been done in the 150 years of ballooning history. A warm and lasting relationship was established with Captain Nordman. Then shortly after the return to Brussels, Van Orman learned that he had been disqualified by the Belgium Aero Club under the rule which prohibits descending into the sea. Interestingly, the race was declared in favor of the Belgian Veenstra, who had not only descended into the sea, but nearly died after seven hours exposure in the sea near Punta Bruntra, Spain. [18]

The Gordon Bennett Race flourished until 1939 when over 1.7 million German troops invaded Poland, crushing the hapless nation and igniting World War II.

The 1939 race was to be held in Poland, the trophy having been won the previous year by Polish Army captain, Anton Janusz. But when the German Panzers rolled across the border late that summer, Polish aeronautical aspirations suffered a double blow—not only was the Gordon Bennett classic canceled, but a high-altitude balloon flight that had been in preparation for nearly two years was also scrapped. Janusz himself would have been the pilot for the attempt on *Explorer II's* altitude record. The United States, under a special export license, granted by the Secretary of State, shipped 220,000 cubic feet of helium from Texas to Poland in July to support the stratospheric attempt. Captain Janusz visited the United States that same month to discuss his attempt with Albert Stevens and to arrange for the helium shipment. When war broke out in Europe before the balloon attempt was ready, the Poles released the helium into the atmosphere to guard against its being captured by the Germans, while Janusz himself, headed for the Polish front. [19]

It was in 1979, after a 41 year hiatus, that a certain atmospheric physicist, Dr. Thomas Heinsheimer, revived the classic Gordon Bennett

Balloon Race. For several years, dual contests were run in the US and Europe. The rules were simple. Competing balloons would be launched from a common place and the winner would be the aerostat that traveled the furthest distance. The winner would take not only the trophy but the honor of hosting next year's event in his/her home nation as was done in earlier years. The competition would not be fully reinstated by the Federation Aeronautique Internationale (FAI) until 1983.

Were there many old fashioned, netted gas balloons available at this time? Not many. In fact very few. Skypower of Tea, South Dakota, would fill the vacuum, at least in part. Skypower produced over seventeen 1,000 cubic meter balloons in the next few years. Skypower also designed a 500 cubic meter balloon, however, this had limited use.

Eighteen balloons were entered into the first revived Gordon Bennett race. Teams that raced with Yost-built balloons included Ben Abruzzo and Maxie Anderson – just a year after their Atlantic crossing; Joe Kittinger and Dewey Reinhard; and Ed Yost himself with Bob Snow. Ed offered to assist in the inflation process for everyone. As each balloon lifted off, a band struck up the national anthem of its pilots: Switzerland, Belgium, Australia, Poland, Italy, Japan, the United Kingdom and the United States. [20]

Abruzzo and Anderson won the race. They landed at Dove Creek, Colorado, a distance of 583 miles from Long Beach, California. Dewey Reinhard and Joe Kittinger took second place while Ed Yost and Bob Snow took third.

We, the gas balloon manufacturers, loved our work at the little Skypower factory in Tea. We were playing an important part in the history of ballooning. I had been a precision draftsman at the Boeing Airplane Company with drawing tolerances — miniscule. When I drafted the blueprints for the 1,000 Cubic meter balloon, Ed would say, "don't sweat the small stuff." We were proud of our work and product. No one was more pleased than the boss, Paul "Ed" Yost.

In 1980, the launch took place at Mile Square Park in Fountain Valley, California. Joe Kittinger and Bob Snow flew in the "poster balloon" which has become the signature Yost-built balloon – *The Rosie O'Grady*. They made second, as did they in the 1981 race. In the 1982 race, Joe crashed near Cody, Wyoming, hitting a fence post and ending up with a dislocated shoulder and broken arm. However Joe won the race traveling 894 miles.

The 1985 Gordon Bennett International Balloon Race got underway Saturday, May 4, in Palm Springs, California, amidst a crowd of about 10,000 spectators who braved sun and heat to witness the historic event. When *Rosie O'Grady,* the winning balloon, finally drifted to earth after nearly 50 hours aloft, Joe Kittinger had made aviation history – again.

Setting an unofficial record for the shortest winning distance ever flown since the Bennett races began in 1906, Kittinger snared his third consecutive victory to retire the coveted Gordon Bennett trophy. It is the first time the perpetual trophy has been retired in modern history of the race.

Kittinger and his co-pilot Sherry Reed, both of Orlando, Florida, flew only 279.9 miles, landing in the desert near Gunderson, Nevada, about 110 miles northwest of Las Vegas. That was enough to win a race in which victory is claimed by the balloon which can fly the greatest distance and land the farthest away from the launching point.

Kittinger described his flight as "extremely frustrating," noting that in his long and colorful ballooning career, he "had never flown so long to go so few miles."

Kittinger gave much of the credit for winning to co-pilot Sherry Reed, who became the first woman ever to win a Gordon Bennett race. It was Reed's first Gordon Bennett as well as her first gas balloon flight. "I said before we launched that it would take two sunsets to win, and it did, " Kittinger said, "Sherry was a tremendous asset to our flight. She was alert and kept watch so that I could get a little sleep. She has never been at this altitude or used oxygen before. At one point we reached 21,000 feet, and I was extremely impressed by her performance." [21]

(Later Joe and Sherry would fall in love and marry.)

At the Survivors' Banquet, held a week later, Sherry Reed accepted the Gordon Bennett trophy for the team, and announced it would be placed on permanent display at Church Street Station in Orlando, Florida. Someone quipped in *Ballooning Life,* "If God wanted women to fly he would have made the sky pink."

Brent Stockwell of Oakland, and team in *Excelsior,* placed seventh, while Ed Yost and Mary Spears of Tea, South Dakota in *Universal,* placed eighth or last.

The first night launch was on May 9 and 10, 1987. Winners were Jim Jones of Chandler, Arizona, and Dale Yost, Columbus, Ohio, with a distance of 49.5 miles.

The record time for the winner of that Gordon Bennett race was held by Germans, Wilhelm Eimers and Bernd Landsmann, who remained airborne for over 92 hours in the 1995 race. The distance record was held by the Belgian duo of Bob Berben and Benoit Simeons, who, in 2005, piloted their balloon 2,100 miles from Albuquerque, New Mexico to Squatic, Quebec, Canada. Austrian Josef Starkbaum won the trophy seven times between 1985 and 1993. American teams have won twelve victories.

The 2010 competition was launched from Bristol, United Kingdom, on September 24. The race was marred by the disappearance of the American team during a storm over the Adriatic on October 1st. The balloon remained missing until December 6th, when a fishing vessel found the cabin (gondola) containing the pilot's bodies off the coast of Italy.

The 2011 race was won by the French team flying *F-PPSE* into Austria. [22]

The aeronauts in the American National Balloon rally and the Gordon Bennett races were men and women with a thirst for adventure. They possessed, as well, a competitive spirit along with great courage. Many succeeded but not all.

In 1923, five competitors were killed when they were struck by lightning while six more were injured in storms. Among the dead were Lieutenants John W. Choptaw and Robert S. Olmstead who were killed when their balloon *US Army S6* crashed in Loosbroek, Netherlands. Sixty years later, in 1983, Americans Maxie Anderson and Don Ida were killed as the gondola detached from their balloon during an attempt to avoid crossing into East German airspace. Anderson and Ida were participating in the "Coupe Charles et Robert" (named for Jacques Charles and the Robert brothers, inventors of the gas balloon) which was run in parallel with the Gordon Bennett Cup. Following Max's and Don's deaths, the Coupe Charles et Robert was never run again. [23]

On September 12, 1995, three gas balloons participating in the race entered Belarusian air space. Despite the fact that competition organizers had informed the Belarusian Government about the race in May, and that flight plans had been filed, a Mil Mi-24B attack helicopter of the

Belarusian Air Force shot down one balloon, killing two American citizens, Alan Fraenckel and John Stuart-Jervis.

A second balloon was forced to land while the third landed safely over two hours after the initial downing. The crews of the two survivor balloons were fined for entering Belarus without a visa and released. Belarus has neither apologized nor offered compensation for the deaths. [24]

On September 29, 2010, the 2004 trophy-winning American team of Richard Abruzzo and Carol Rymer Davis went missing in thunderstorms over the Adriatic Sea. The balloons cabin containing the bodies were recovered by an Italian fishing boat on December 6th. [25]

Balloons Aloft connects the International Gordon Bennett Race, in this brief overview, to the role South Dakota has played in the promotion, participation and celebration of this amazing sport of gas ballooning.

Chapter Sixteen

Crossing the Big Pond

"We cannot discover new oceans unless we have the courage to lose sight of the shore." – Andre' Gide

A young man, eyes bright with excitement, stepped into the Skypower-Universal Systems factory one day, asking for the man who makes balloons. "I want to cross the Big Pond," he announced. The "Balloon Man," Ed Yost, appeared from the back, greeted him, and without hesitation laid out the costs, the dangers and the failures of the past of those who attempted to "cross the Atlantic by balloon." The prospective young adventurer realized his fanciful dreams lay elsewhere – "crossing the pond" would be someone else's glory.

> *"It has occurred to me, sometimes, late at night,"* confesses one balloonist, *when my head is not quite clear, when I've had too little sleep and too much champagne, that I don't want to fly a balloon at all. I want to be a balloon."* —The Ballooning of America

The following story appeared in the New York Sun in 1844:
Astounding News! By Express via Norfolk: The Atlantic Crossed in Three days! Signal triumph of Mr. Monck Mason's Flying Machine!!! Arrived at Sullivan's Island near Charlestown, S. C., Of Mr. Mason, Mr. Robert Harrison Ainsworth and four others in the Steering Balloon

"*Victoria,*" after a passage Seventy-Five Hours from land to land. Full Particulars of the Voyage!!!"

Many papers were sold, but the *New York Sun Time's* story was a hoax. The Atlantic crossing story was retracted two days later. An impoverished writer, Edgar Allen Poe, had written it, having arrived in New York the preceding week. Poe's wife was sick. He was broke. The tale was a desperate attempt to make money. "The newsmen called, 'balloonantics,' had been royally spoofed and everyone had a good laugh – everyone, that is, except a certain young Thaddeus S. C. Lowe." [1]

Thaddeus Sobieski Constantine Lowe (1832-1913) from Jefferson, New Hampshire, dreamed of crossing the Atlantic Ocean by balloon. He was the most distinguished American aeronaut in the nineteenth century. Lowe gave full credence to John Wise's *Upper Wind System Theory.* Today we call it the *Jet Stream* phenomenon. On September 8th 1860, Lowe began to inflate the balloon he expected to carry him across the Atlantic. The balloon, the *Great Western,* was 200 feet high and 104 feet wide and had a capacity of 725,000 cubic feet. It was so big that in the previous year, at the original inflation site at the corner of Fifth Avenue and 42nd Street in Manhattan, it drained all the gas from the New York Gas Company, and the inflation had to be canceled. This time in Philadelphia, the balloon simply burst.

April 19, 1861, T.S.C. Lowe attempted a second transatlantic balloon flight. Lowe used a smaller balloon than the *Great Western,* and launched a test flight to the coast. The flight went perfectly for 10 hours, however, trouble appeared when he began to descend. In South Carolina he was surrounded by a hostile crowd. When the locals threatened to lynch him, Lowe reached for his Colt revolver, and the Southerner crowd backed off. He saved his life, but ended up spending the night in a nearby jail under the pretense that he was a Yankee spy. Lowe would in fact use his balloons and expertise in the Civil War effort for the North. He was appointed by President Abraham Lincoln, Chief Officer of the Union Army's first Aeronautic Corps.

John Wise, a professional balloonist from Lancaster, Pennsylvania, studied the air currents for meteorological purposes and learned that upper air currents in the Midwest blew from west to east. He was convinced that he could carry light mail and passengers in a balloon from the Midwest to the East Coast. John Wise believed it was also possible to fly across the Atlantic. In March of 1873, Washington H. Donaldson, a balloon daredevil, collaborated with John Wise and the newspaper *The Daily Graphic.* Wise departed company with Donaldson after a failed balloon Atlantic crossing attempt. "However, on September 10, 1873, a transatlantic crossing was attempted with Professor W. H. Donaldson and two other passengers." [2] When the wind whipped the envelope and threatened to fly away, the attempt was aborted.

On October 7, 1873, another attempt was made. The launch went well until the aerostat encountered a thunder storm. The *New Graphic* plummeted. The aeronauts bailed out. The attempt simply ended up in the history books as the first real attempt at crossing the Atlantic expanse.

Would crossing "the Big Pond" ever happen? Was flying the Atlantic in a gas balloon simply a fanciful adventure that would elude even the most courageous?

Nonetheless, the dream of conquering the ocean, persisted in the imaginations of modern 20th century balloonists. Improved plastic envelopes, radios, and new technological devices brought a transatlantic voyage temptingly within reach. Of course such a flight would always require skill, risk and luck.

Enter **Arnold Eiloart** and **Colin Mudie,** veteran English yachtsmen, who had sailed by boat across the Atlantic, but had never even ridden in a balloon. They decided to tackle the ocean by air. After certification, they designed *Small World,* a 53,000-cubic-foot hydrogen aerostat with a seaworthy gondola that could be converted to a sailboat if circumstances required it. On December 12, 1958, Arnold's son, **Tim, Colin,** and wife **Rosemary,** lifted off from Tenerife, Canary Islands. They ascended in a rising gale and quickly soared westward. The team carried a drag rope, invented by fellow countryman Charles Green, but added something new— a water bucket with which to scoop up sea water at the end of a 3,000 foot line. "On the fourth night the *Small World,* ran into vigorous updrafts until it resembled, as Eiloart recalled, a "ride in an elevator controlled by a mischievous child." *Small World,* its ballast all but gone, ditched at sea after 1,200 miles. Their duration was 94.5 hours. They sailed to safety in their sail-equipped gondola-boat.

After the *Small World's* failure, more than a decade would elapse before anyone would try to brave the mighty Atlantic Ocean.

August 10, 1968, **Mark Winters** and **Jerry Kostar** launched from Halifax, Nova Scotia, in a 35,000 cubic foot helium balloon–*Maple Leaf*. The duo traveled 70 miles in 20 hours and landed 35 miles from Halifax. They were rescued.

The next flight, with three aeronauts, was tragically lost at sea. **Malcolm Brighton, Rodney Anderson and Pamela Anderson** flew the balloon *Free Life*, which was made up of a helium and hot air combination of 73,000 cubic feet. The trio traveled, it is believed, about 1,400 miles in 30 hours. They were lost somewhere east of Newfoundland.

Bob Sparks flew his *Yankee Zepher*, August 7, 1973, from Bar Harbor, Maine. His helium and hot air combination balloon carried him 850 miles in 23.5 hours. He landed 45 miles north east of St. Johns, and was rescued.

Tom Gatch, in a balloon called *Light Heart,* lifted off from Harrisburg, Pennsylvania, February 18, 1974. Tom's balloon utilized a helium composite and unfortunately after 1,400 miles, plus or minus, and 18 hours in the air, was lost at sea.

In a helium balloon, *Spirit of Man,* on August 6, 1974, **Bob Berger** traveled only 12 miles in 1 hour landing at Barnegat Bay, New Jersey. He launched from Lakehurst, New Jersey.

January 6, 1975, **Malcolm Forbes** and **Tom Heinsheimer** lifted off in *Windborne* from Santa Ana, California. The flight was aborted so no distance nor duration records were set.

Bob Sparks tried again with **Haddon Wood,** a stowaway, launching August 21, 1975, in another hybrid balloon consisting of 100,000 Cubic feet. The *Odessey* lifted off at Mashpee, Massachusetts, flew for 125 miles in 18 hours, landing 125 miles south of Cape Cod in 12 foot seas. The Canadian Coast Guard rescued them.

Karl Thomas, in the *Spirit of 76,* a 77,000 Cubic foot balloon launched from Lakehurst, New Jersey, traveled 550 miles in 33 hours and was rescued 375 miles NE of Bermuda. [3]

One man, who knew as much or more than any balloonist, was keeping his eye on the sky and a newspaper in hand. Ed Yost at Tea, South Dakota said, "I kept reading these articles, with people making their balloon, flying them, and going out 100 or 300 miles or so, then landing in the ocean, so I thought, 'hey, it isn't that big a deal.'" [4]

The crew at Skypower went to work. Bang, swish, wurrr, zip, slosh! A balloon adventure always begins with the aerostat's design and construction. The neoprene coated nylon bag (envelope) would be relatively small for this trip–60,000 cubic feet of helium. A test inflation showed the envelope's shape and snout-like appendix, which would allow for the release of helium when expanded by heat from the sun's rays. This expansion effect would be lessened by a coat of reflective silver paint on the upper hemisphere of the balloon. The lower part would be painted black to absorb heat. What would the name be? It would be *Silver Fox*.

The gondola was constructed as a safety measure into a catamaran boat for possible landing at sea. The boat idea came from Arnold Eiloart and Colin Maudie, the yachtsmen who attempted the Atlantic crossing in 1958. The gondola was painted a bright red so it was clearly visible from sky and sea. Instruments included 2 barographs for proof of continued flight of the *Silver Fox*. The instrument panel was crammed with sensitive navigational and radio equipment. To avoid touchdown at sea while Ed, the pilot, was sleeping, a barometer was linked to a motorcycle horn that would shrill a warning should the balloon dip below 3,000 feet. Food, water, cameras, ballast–the total weight would be less than 2 tons. More than half of the 2 tons would be ballast and expendable equipment. Weather, maps and projections were studied carefully for the 3,000 mile trip across the "Big Pond."

The sponsor for this trip was the National Geographic

The *Silver Fox* flies over the Gulf of St. Lawrence on it's way to Europe.
–Photo by Otis Imboden/National Geographic Stock

Society. The adventure would be written by Ed and published in the National Geographic Magazine, February 1977 issue.

The crew loaded Ed's vehicle – the *White Elephant* – and headed out from South Dakota to Milbridge, Maine. At the launch site, the envelope was filled with helium at a cost of between 2 ½ to 4 cents per cubic foot. Ed climbed aboard, waved goodbye and flew off for Europe.

On the first day, the *Silver Fox* caught a gust of wind and drifted slowly northward at 5,000 feet above Canada's Gulf of St. Lawrence. Silence and seas surrounded Ed midway across the ocean. On the third day the he drifted at an altitude of 9,660 feet. Below were small cumulus clouds – overhead a layer of cirrus clouds which would bring new problems. The cirrus would screen the sun, thus reducing the warming of the helium and would retard the buoyancy that is needed. Precious ballast was heaved overboard in order to maintain altitude. [5]

On the next day Ed was being driven irresistibly toward the Azores. He was losing altitude. He threw everything he could overboard..canned food..sleeping bag...radios.... receivers.... the sextant. He would have to crash. He valved and valved and then pulled the rip cord. He touched down, landing in the sea near the West German freighter *Elizabeth Bolten*. He was 700 miles off the coast of Portugal.

A lockheed HC-130 rescue plane hovered overhead. It was from the U.S. Air Forces' 67th Aerospace Rescue and Recovery Squadron based at Wood Bridge, England. "We're going to drop you a book to read," the HC-130 commander radioed. "What kind of book?" Ed asked. "Jaws" came the cheerful reply.

Upon Ed's arrival in Gibraltor he received a cable from President Gerald Ford which said, "YOU SHOULD TAKE GREAT PRIDE IN YOUR TRULY OUTSTANDING ACHIEVEMENT AND I EXTEND MY VERY BEST WISHES TO YOU, THE MEMBERS OF YOUR FAMILY AND TO THE PEOPLE WHO HELPED MAKE THIS FEAT POSSIBLE." [6]

Ed Yost, through skillful use of ballast and valving, had maneuvered for a day and a night between 5,000 and 14,600 feet in a vain search for the winds that both poet, scientist and old John Wise said were there in the upper air. Ed flew 107 hours, 37 minutes, traveling 2,740 miles, breaking all distance and duration records. In 1913, German balloonists, Hugo Klen and Alfred Schmitz, had stayed aloft for 87 hours, setting a duration record. In 1914, Hans Berliner and Alexander Haase flew 3,053 km (1832 miles) from Bitterfield, Germany to Perm, Russia, a distance record. Those two class records stood for 62 years until Ed Yost exceeded them both during his historic flight. If his flight was a failure at crossing the Atlantic, it was at the very least a noble and successful failure.

After breaking the Germans' records, Ed was asked if he would try it again, and he replied with an unequivocal "NO!" Ed didn't try it again but played significant roles in the ones to follow.

The late and famous balloonist Charles Dollfus, known as the premier European balloonist and balloon curator, congratulated Ed with this letter dated July 5, 1976 and sent from 82, Rue du Ranelagh Paris: *"Dear Edward, I received with a shock of interest and friendship your letter with the news of your Atlantic attempt. You know that in all my long air life, I was and am interested in all which concerns the Ocean flying and air crossing, and of course, especially the Atlantic which, by the way I practiced in the 30 years in airship. I perfectly feel that you keep for yourself your project and special works, but it if is not to take too much of your time, and if some information is available, will you be so kind as to send me as much as possible. I had some good ascents, this year, the last one in Balleroy.*

I heartily and often think of you. Yours very friendly, (signature) Charles Dollfus, Aeronaut."

Successful business man Max Anderson, 42, sleepless one night, opened a copy of the National Geographic and read the story of Ed Yost's attempted flight across the Atlantic. He was fascinated. He called his old friend Ben Abruzzo, who was also an enthusiastic hot air balloonist. "What would you think about you and me flying the Atlantic?" Ben replied, "Let's do it."

It was not long before a night in Albuquerque, New Mexico, when two adventurist, wealthy businessmen met to discuss "crossing the Big Pond." They were Ben Abruzzo and Maxie Anderson. They knew, "In 1977, only one man in the United States was in the business of building transatlantic balloons – Ed Yost. Ben and Max went up to see him in April of that year, after some preliminary talk on the telephone. Max was negotiating at the time for the purchase of a new company airplane, and the dealer, Bill Cutter of Albuquerque, offered to fly him and Ben up to Yost's factory near Sioux Falls in a new Beechcraft King Air.

Ben and Max arrived in Sioux Falls in time for dinner. The world of ballooning is a quiet world and Ed Yost is a quiet man—-'a man of few words and those well chosen,' as Max was fond of saying. Ed Yost conducted them to a restaurant that featured a harpist and two young women playing violins. They had seafood for dinner. This was their first prolonged meeting with Yost. That evening at his house, Yost showed Ben and Max his memorabilia – color slides of the flight of the *Silver Fox;* a cable from President Ford, sent on the occasion of Yost's splashdown in the Atlantic next to a West German freighter...Max was to remember that combination of images for a long time afterward." [7]

After this meeting Ben and Max decided to have Yost build a balloon to cross the Atlantic.

"'The price,' said Yost, 'would be $50,000 for the balloon.' Ben tried as any business man would to negotiate with Ed. Yost simply repeated the figures: that was his price; take it or leave it. 'Deal,' said Ben to Ed Yost. The three shook hands." [8]

The new balloon would be called *Double Eagle* named after Lindbergh's airplane flight – the *Lone Eagle,* which crossed the Atlantic, from New York to Paris, solo and nonstop in 34 hours. The year was 1927.

However, in 1977, two teams were preparing to attempt the crossing. In Colorado Springs, it was Dewey Reinhard and Steve Stephenson now competing with Max Anderson and Ben Abruzzo. Ed Yost was building balloons for both.

"'The decision to have Yost construct his balloon was never in question,' according to Dewey Reinhard. 'I visited with all of the manufacturers and his had the best track record, set on his own attempt....He (Yost) taught me to fly gas balloons in Amarillo, Texas,' says Reinhard. 'At the time, I really didn't think he was too good an instructor. It wasn't until later, and I guess it's like that with a lot of things, that you realize how much you absorb. When we took off from Bar Harbor on our attempt that was the first time, I'd never been in a gas balloon for an extended period without Ed along. We had some stormy times. One minute Ed can be a raving mad individual and the next moment be a gentle, caring person.... there is no one else I would trust my life to more than Ed Yost.'" [9]

A third party had contracted with Ed Yost to build a transatlantic balloon. This would be a solo attempt by Joe Kittinger. The balloon and gondola for Joe were built, but the plan was scrapped when expected sponsorship bailed.

Dewey Reinhard and Charles Stephenson would launch in the Yost built balloon *Eagle,* filled with 86,000 cubic feet of helium, on October 10, 1977 at Bar Harbor, Maine. They traveled 200 miles in 46 hours and had to be rescued 50 miles SE of Halifax. However before their liftoff, Ben and Max in *Double Eagle* had launched from Marshfield, Massachusetts, September 9, 1977. The *Double Eagle* was designed with a 101,000 Cubic Feet envelope. Ben Abruzzo and Maxie Anderson had nothing but trouble immediately. A tardy takeoff pulled them into strong downdrafts over the Gaspe Peninsula in eastern Canada. They were sucked down to within 100 feet of the tree tops. They threw overboard almost one third of their ballast. They struggled with sleet and snow and rain. By the time *Double Eagle* had been in the air for 36 hours, Ben Abruzzo was sure that he was freezing to death. They radioed for a United States Air Force rescue helicopter from Woodbridge, England and ditched. They were rescued, 3 miles from Northwest Iceland, but in bad physical shape.

The night before their 2nd historic flight, the 3 balloonists were introduced to singer Jeanie C. Riley (*Harper Valley PTA* fame) who was appearing at the Northern Maine fair. (Left to right) Anderson, Newman, Abruzzo, Riley.
– *Voscar, the Maine Photographer, Presque Isle, Maine*

Once more, the mighty Atlantic had won, even though the courageous duo had flown 2,440 miles in 66 hours.

Within 24 hours after the ditching of *Double Eagle*, Max Anderson said he knew he was going to try again. The original 17' x 6 1/2' gondola had been retrieved from the sea to be used again. Fitted with a more effective rain shield and rigged with a shelter containing a propane heater, the gondola would be much more comfortable. Wool clothing and oilskins, as well as electrically heated socks, would be used.

August 10, 1978, *Double Eagle II* took off from Presque, Isle, Maine. This launching site would shorten the starting point by 300 miles. The balloon was substantially larger than *Double Eagle*.

A third crew member, Larry Newman, a hang-glider enthusiast, had been added to the crew. A three man crew would enable one to sleep, one to rest, and the other to stand watch. The book *Double Eagle* by Charles McCarry, describes a fascinating saga into the temperaments and arguments that went on during the flight. Ed Yost told me that Max and Ben even got into a fist fight. Max did not care for Larry, whereas Ben had been the one who selected Larry.

The *Double Eagle* II would surf toward Europe ahead of a menacing storm front, just as predicted by meteorologist Bob Rice. It appeared some of the earlier problems might just be overcome. Before launch the helium

Double Eagle II is ready to launch at Presque Isle, Maine. Note Newman's hang glider on the right.
– *Voscar, the Maine Photographer, Presque Isle, Maine*

supplier, trying to save the balloonists money, sent only one truckload of the gas, less than had been ordered. The inflation, supervised by Ed Yost, went too slowly, and the aeronauts took off with an envelope only 85% filled. Even though it was a late launch, the flight turned out to be a magnificent and successful adventure! On August 17, at 10:02, the trio crossed the Irish coastline and into the history books! At dawn *Double Eagle II* came in sight of Wales. **The Big Pond had been crossed!**

A squadron of small airplanes and helicopters appeared as they crossed the English Channel. Over southern England, the aeronauts were dumbfounded to see the morning sun sparkling from thousands of hand-held mirrors, as a signal of welcome from the throngs below. A message from the Smithsonian Institution in Washington, D.C., was relayed by radio from London. Would the balloonists be willing to donate *Double Eagle II* to the Smithsonian? The aeronauts replied they would be honored. Abruzzo desperately hoped to land in Paris, preferably at Le Bourget Airfield, where Lindbergh had touched down in 1927. But instead *Double Eagle II* landed in a barley field next to a highway just outside the village of Evreux, 50 miles west of Paris.

The *Double Eagle II* flight was the climax of one of ballooning's finest achievements in a nostalgic return to its French birthplace.

Six years later, history would be made again. This time it would be Colonel Joe Kittinger in a South Dakota Yost-built gas balloon. It happened this way:

"In the spring of 1984, a Canadian businessman named Gaetan Croteau contacted Ed Yost about his desire to sponsor a record-setting balloon flight. He was willing to spend up to $250,000. Ed told him, 'I know just the guy for you,' and in a matter of weeks I (Joe) was at work once again on a transatlantic flight. But this time there were no ifs, ands, or buts. It was to be a *solo* flight. I made sure that was understood.

Ed fortunately still had the balloon and the gondola he'd built for me back in 1978. There were some modifications to be made, but we weren't too far from being ready to go. Once we had the money we needed in the bank, I called Bob Rice in Massachusetts. He was not only an ace meteorologist, but he knew aeonautics. A tremendous talent." 10

South Dakota history-making ballooning would make headlines on September 18, 1984, when Colonel Joseph Kittinger revived and lived his "Hanoi Hilton" dream, if not of flying around the world in a balloon, but of at least crossing the Atlantic solo.

Plans were made. The Canadians required a flight plan since Canada is responsible for air and sea rescue in the western half of the northern Atlantic Ocean. Joe's flight plan said simply:

POINT OF DEPARTURE: *Caribou, Maine*
DESTINATION: *Unknown*
ROUTE: *Unknown*
DURATION OF FLIGHT: *Unknown*
ALTITUDE:*Unknown*
FUEL ON BOARD: *Zero*

"Having the paperwork seemed to make the Canadians happy." [11]

Joe and crew were ready to launch as soon as meteorologist Bob Rice gave the word. He gave it. "In my effort to cover all my own bases, I had invited a Catholic priest and a Baptist minister to attend the launch. I'd tried to find a rabbi but hadn't been able to come up with one up in Caribou. If I'd known about a Buddhist in the area, he would have been invited. We assembled the team and the townspeople who'd come out to witness the event, and after a couple of prayers we all drank a champagne toast to a successful flight. I gave Sherry a big kiss, and I climbed into Ed's gondola. In honor of Bob Snow, we called it the *Rosie O'Grady*. The balloon looked great, and everyone was in high spirits.

Joe Kittinger is about to launch from Caribou, Maine in the *Balloon of Peace- Rosie O'Grady*.
—*Voscar, The Maine Photographer, Presque Isle, Maine*

At about 8:00 p. m., I waved and weighed off. Away I went into the black of night. It was really a magnificent moment for me. I'd been working for this for more than ten years. From the Hanoi Hilton to the coast of Maine." 12

Joe Kittinger made it across the Atlantic Ocean in 83 hours and 40 minutes. He landed by crashing through tree tops and on a rock, breaking his foot as he tumbled out of the gondola. The place was Cairo, Montenotte, Italy. Joe had traveled 3,543 miles, surpassing the Anderson, Abruzzo, Newman flight, by 500 miles. This would be a record likely to stand for a long time.

Joe and Sherry celebrate Joe's successful flight.
— *Voscar, The Maine Photographer, Presque Isle, Maine*

Joe was celebrated as a hero not only by the Italian government but also by the citizenry of the United States and not least of all, the folks who puttered away in a little factory in Tea, South Dakota.

Joe presented the flag flown on the *Rosie O'Grady* to President Reagan. Joe appeared on The David Letterman Show and received innumerable honors.

Twice, Crossing the Big Pond, had been accomplished.

Chapter Seventeen

Awards, Accolades, and Accomplishments

"May the Winds Welcome you with softness
May the Sun bless you with his warm hands
May you fly so high and so well
That God joins you in laughter

And may God set you gently back again
Into the loving arms of Mother Earth."

– A Pilot's prayer believed to have
originated in Ireland

The *Hastings Tribune* called it "A Hero's Welcome." An assortment of folks gathered for an historic and festive event, honoring Ed Yost at the abandoned Army Air Field near Bruning, Nebraska. The event, May 21, 2001, witnessed Ed and Suzanne Yost cutting a red ribbon and dedicating a granite marker set along Nebraska Highway 4, seven miles east of Bruning. The granite monument recognized Ed's first flight in a modern hot air balloon. The granite stone, set against a backdrop of green fields and blue sky, related how at that place, history was made October, 1960. Ed, Jim Winker and other crew members from the Raven Industries balloon division had test-flown a new hot air invention on the runway that would one day revolutionize sport ballooning in South Dakota, and throughout the world. Sponsors for this event were the Outhaus Balloon Club and the Nebraska State Historical Society. Donations from the ballooning community across the United States, and from England made the marker possible. Peggy Hart of Campbell, Nebraska, a member of the Outhaus Balloon Club, said, "its a gift to show how they feel for you. They think it was about time you (Ed Yost) were honored." Ninety pilots, families and local residents turned out for the honoring event in Bruning. They come from sixteen states from California to Connecticut to Florida.

The weekend started with nine balloons taking off Saturday morning from the Bruning football field. A trip to the Thayer County Museum in Belvidere by the assembled group witnessed Ed's donation of a framed picture of the first flight to the museum. A banquet in Ed's honor followed at the Bruning Opera House. Joe Kittinger, longtime friend, introduced Ed saying, "We're here to honor a real pioneer, a defender of his country around the world who took part in many classified programs for the U. S. Military." At the luncheon at the Opera House, Ed spoke of his career through modest recollections and a mix of humorous stories. He jokingly called himself a member of "The Plutonium Sand and Gravel Company" since he performed many classified programs for the government. "Everybody wants to know why I picked Bruning," he said. He explained that he had flown over the former Army Air Field several times en route from the Raven factory in Sioux Falls, South Dakota to Texas. "One main reason for choosing Bruning, was its isolation and unobstructed places," he explained. At the ribbon cutting ceremony, Ed mumbled a modest joke about how normally one doesn't get such an honor when he is still alive. Ed was 81 at the time.

The roadside marker reads:

FIRST FREE FLIGHT
OF A MODERN HOT AIR BALLOON
ON OCTOBER 22, 1960, PAUL E. "ED" YOST MADE THE FIRST FREE FLIGHT OF A MODERN HOT AIR BALLOON FROM THE FORMER BRUNING ARMY AIR FIELD. THE FLIGHT LASTED 23 MINUTES AND COVERED 3 MILES.
HUMANS FIRST FLEW IN 1783 IN A HOT AIR BALLOON, CONCEIVED AND BUILT BY THE MONTGOLFIER BROTHERS IN FRANCE 120 YEARS BEFORE AN AIRPLANE FLEW. THE

HOT AIR BALLOON LANGUISHED UNTIL 1956 WHEN ED YOST RECEIVED A CONTRACT FROM THE OFFICE OF NAVAL RESEARCH TO DEVELOP A PRACTICAL HOT AIR BALLOON. YOST DESIGNED THE SHAPE OF THE ENVELOPE, PIONEERED THE USE OF NYLON AND DEVELOPED PROPANE HEATERS THAT MADE CONTROLLED AND SUSTAINED FLIGHT POSSIBLE. WHILE WORKING ON SCIENTIFIC PROJECTS YOST MADE MANY AIRPLANE FLIGHTS FROM RAVEN INDUSTRIES IN SIOUX FALLS, SOUTH DAKOTA TO TEXAS FLYING DIRECTLY OVER BRUNING ARMY AIR FIELD. AFTER LAUNCHING 2 LARGE HELIUM SCIENTIFIC BALLOONS FROM BRUNING, YOST DECIDED IT WOULD BE A GOOD LOCATION FROM WHICH TO TEST FLY THE NEW HOT AIR BALLOON BECAUSE THERE WAS OPEN SPACE AND FEW OBSTRUCTIONS. ED YOST'S MODERN HOT AIR BALLOON FIRST FLOWN IN BRUNING, TRANSFORMED BALLOONING FROM AN ELITE ACTIVITY TO A SPORT OPEN TO THOUSANDS OF BALLOONISTS WORLDWIDE.

Monument notes the first free flight of a modern hot air balloon near Bruning, Nebraska.

Host Peggy Hart noted, "those Bruning test flights opened the way for the sport of hot air balloon. Prior to that, the sport was for the elite. Balloons were powered by helium which was expensive; hydrogen which was volatile; or smoke, which was less than ideal." [1]

Ms. Hart and the other planners, pleased by the local turnout, was ecstatic by the overwhelming response of the ballooning community nationwide. The monument sits next to an existing historical marker commemorating the former Bruning Army Air Field.

On August 4, 2002, two genuine American heroes were honored by the residents of Bristow, Iowa, with the dedication of two granite monuments. The two heroes were Ed Yost and Col. Joe Kittinger Jr. One monument was dedicated to all veterans with the words: IN MEMORY OF THOSE WHO SERVED. WE ARE PROUD TO HONOR ALL VETERANS IN AND AROUND BRISTOW WHO SERVED OUR COUNTRY. And the second monument was dedicated to Ed Yost with the words: BRISTOW IS PROUD TO BE THE HOMETOWN OF PAUL "ED" YOST FATHER OF THE MODERN HOT AIR BALLOON AND AVIATOR. A 3-volley gun salute by the Honor Guards of Allison, Iowa, the Bristow Amvets, and the Dumont American Legion, was followed by a haunting but powerful rendition of taps. My wife Pam and I (Arley Fadness) attended and thoroughly enjoyed the dedication ceremony as well as the events of the 3 day celebration. "Bristow's 'Veterans Day' began at 5 a. m. August 2nd with Ed Yost playing a prank on the citizenry. He arrived from New Mexico with his friends, Beth and Bill Gibson and others, in a caravan honking his horn up and down the narrow streets of this tiny spot in northeastern Iowa. He said to 'drive through town honking to wake everybody up, so we'll have an audience out there.' Several vehicles followed suit. Their horns blared, letting everyone know the father of the modern hot air balloon was in town and it was time to howl. Get up and out to the launch field and watch 'em fly." [2]

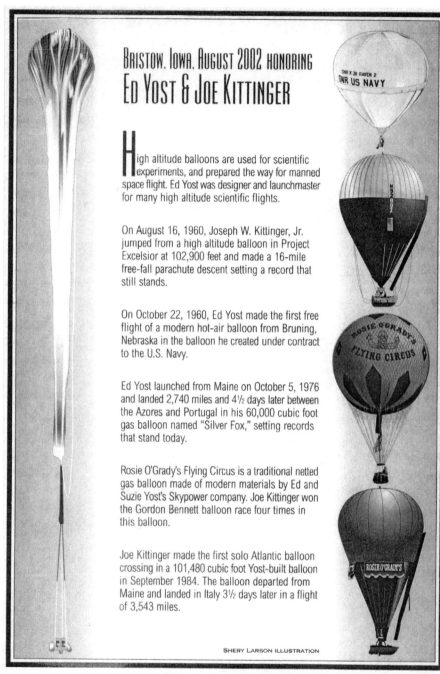

Bristow, Iowa, August 2002 honoring
Ed Yost & Joe Kittinger

High altitude balloons are used for scientific experiments, and prepared the way for manned space flight. Ed Yost was designer and launchmaster for many high altitude scientific flights.

On August 16, 1960, Joseph W. Kittinger, Jr. jumped from a high altitude balloon in Project Excelsior at 102,900 feet and made a 16-mile free-fall parachute descent setting a record that still stands.

On October 22, 1960, Ed Yost made the first free flight of a modern hot-air balloon from Bruning, Nebraska in the balloon he created under contract to the U.S. Navy.

Ed Yost launched from Maine on October 5, 1976 and landed 2,740 miles and 4½ days later between the Azores and Portugal in his 60,000 cubic foot gas balloon named "Silver Fox," setting records that stand today.

Rosie O'Grady's Flying Circus is a traditional netted gas balloon made of modern materials by Ed and Suzie Yost's Skypower company. Joe Kittinger won the Gordon Bennett balloon race four times in this balloon.

Joe Kittinger made the first solo Atlantic balloon crossing in a 101,480 cubic foot Yost-built balloon in September 1984. The balloon departed from Maine and landed in Italy 3½ days later in a flight of 3,543 miles.

SHERY LARSON ILLUSTRATION

– Photo courtesy of Shery Larson

196

Ed Yost had agreed with Bristow Mayor Karen Cornwell to partici-
pate in the celebration if his good friend Joe Kittinger Jr. would also be
honored. At the Friday evening dinner a replica of the Bruning historical
monument, made by the members of the Outhaus Balloon Club, was
presented to Ed. Ed's family came from many places to honor their
father, grandfather, uncle, and cousin. Ed's oldest son Dale flew in from
Singapore, son Greg drove up from Louisiana, granddaughter Nicole
flew from San Diego, and many relatives and in-laws from Iowa and
Minnesota gathered to honor their relative and friend. It seemed as if
every third person was a Yost by birth or marriage. [3]

Limited Edition. Balloon mail flown at Bristow, Iowa honors
Ed Yost, Joe Kittinger and Clayton Folkerts. — *Photo courtesy of Shery Larson*

The Bristow committee created a piece of balloon mail to commemorate this special day. Shery Larson, artist and balloonist, designed the
envelope. Each piece of mail was signed by Ed and Joe.

Saturday morning, the day to carry the official mail, was ominous at
the beginning but turned out close to perfect. Seventeen balloons flew,
while Ed and his dog Shadow greeted all the pilots and crews.

Whereas Suzie Yost, Ed's wife, friend and balloon partner, had died the year before – her ashes were buried during the Bristow celebration on Saturday, August 3rd in the Yost family plot in Allison, Iowa. Pastor Arley Fadness, of Custer, South Dakota presided at the graveside committal service. Several hundred people said good bye to Suzie, in that quiet Iowa countryside bathed by a shining sun. Earlier on September 15, 2001, in the workshop where Suzanne Robinson Yost had built balloons, a few friends who respected and loved her gathered for a time to celebrate her life. Suzanne was born in Los Angeles in 1932 and died September 2, 2001. She died the day after 911. So after the terrorist's attack, there were many complications for air flight throughout the country. Joe Kittinger, Ed and two of Suzie's long-time friends and colleagues were flying from Florida to Denver on that September 11th, when the pilot announced that they had been ordered to land at Atlanta. The four had to wait a day before they could rent a car and drive to New Mexico. Others had difficulty in getting to the Vadito cabin and balloon shop now transformed into a chapel, such as Dale, Ed's son who lived in Singapore. Larry and Karen Conwell, the mayor of Bristow, Iowa, Becky Pope from Minnesota and Duane and Mary Powers from California who were able to drive.

In the shop, a dais was made from a Skypower balloon basket. Someone played a small organ while a local pastor welcomed and spoke of Suzie's life. Joe Kittinger spoke, reminding all that she had been nicknamed "Spider Woman" because of the balloon nets she made. Each net had over 40,000 knots. I, as the drafter, had worked closely with Suzie drawing each knot to scale. Suzie had been the communicator, liason, mediator, translator and the grace in the Skypower shop. Most all correspondence to me about the projects were done for Ed through Suzie.

Christine Kalakuka, in her laudatory piece entitled, *Celebrating a Good Life: Suzanne Robinson Yost,* described her life. "Suzie was part of a small group of Southern California friends who bought the first sport balloon sold by a small company called Raven Industries. In 1962, Bill McGrath, dean of students at USC, Dick Higbie, lawyer, and Suzanne Robinson, law clerk, learned to fly their balloon in the desert near Palmdale, and then in Mexico after the Palmdale fire marshall got nervous.

Suzie worked for the law firm of Higbie and Higbie as a clerk. She decided she would rather be a lawyer. No law school for Suzie — the lawyers in the firm were her teachers; she studied very hard, passed the bar exam on the first try and joined the firm. Later she became a judge.

A balloon race from Catalina, Island, to the mainland in January, 1964, was to be pivotal in Suzanne Robinson's life—although she wouldn't know it for almost 15 years—for that is when she met Ed Yost. Ed was one of the participants in the race; not only was he the winner, his balloon was the only one to reach the mainland. Ms. Robinson was his assigned crew and with one other person helping, she and Ed, in pouring rain, packed up his soggy balloon from the beach at Dana Point.

Ed and Suzie were married in 1978. They formed Skypower Corporation and were very instrumental in the rebirth of gas ballooning that began with Tom Heinsheimer's resumption of the

Gordon Bennett Race in 1979. Together Ed, Susie and crew built 17 Skypower balloons for the GBBR.

Craig Ryan, author of *The Pre-Astronauts,* wrote in a letter that was read as part of Suzie's celebration: "she seemed to have it all: beauty, smarts, grace, humor, kindness and a tough shrewdness—you could see it in her eyes—that made it clear she wasn't to be trifled with. We all admired the courage and optimism with which she faced what she knew would be the end. I know that when my own day comes, I will wish for the counsel of someone like Suzie Yost. She was just a terrific, marvelous person. I'm honored to have known her, and I can promise you that I'll never forget her. " [4]

Suzanne was Episcopalian; I, a Lutheran Pastor. We often discussed and joked about the faith as we saw it. Once, when Suzanne and Ed were in Europe, she sent a card to me which was a photo of a cathedral they had visited. She wrote, "This picture of the cathedral recently struck by lightning –burned out the entire roof. Some speculate it was because one of the clergy didn't believe 100% in the Apostle's Creed." Suzie was truly a lady of grace, with a quick wit, and engaging smile.

I smiled when I received a photo of Ed and Suzie's lovely mountain home in the Sangre de Cristo mountains of New Mexico. To build their house, Ed and Suzanne used a hot air balloon to lift supplies to the top of the roof. Who else, but Suzie and Ed? I too, promise I will not forget her. Let this text be, at least, a partial fulfillment of that promise.

Ed Yost and Joe Kittinger received many awards and worthy recognitions for their contributions and careers. Ed was the co-founder of the Balloon Federation of American with Don Kersten and Peter Pellegrino. He received the Montgolfier Diplome in 1976. On Friday, June 11, 1999, Ed became the 48th person to receive the distinguished Godfrey L. Cabot Award from the Aero Club of New England. This highly prestigious Cabot Award, named for the founder of the Aero Club of New England and the first American president of the Federation Aeronautique Internationale (FAI), had been awarded to giants of aviation including Igor Sikorsky, General Curtis Lemay, Dr. Charles Stark Draper, General James Doolittle, the Rutan/Yeager Voyager team, jet engine inventors, Sir Frank Whittle and Dr. Hans Von Ohain, Alan Shepard Jr., Walter Schirra Jr., Barry Goldwater, and Col. Joe Kittinger Jr.

Ed received awards from the Wingfoot Lighter-than-air Society, the National Aeronautic Association as the NAA Elder Statesman, and the U. S. Ballooning Hall of Fame. He traveled to England with granddaughter Nicole to receive the prestigious Lipton Trophy by the British Balloon and Airship Club in 2006.

STATE OF NEW MEXICO **EXECUTIVE OFFICE** SANTA FE, NEW MEXICO

Proclamation

WHEREAS, *Governor Bill Richardson and International Young Aeronauts recognize the importance of the unique and significant contributions to our state and country by Aviation Pioneer, Mr. Paul Edward Yost; and*

WHEREAS, *Mr. Yost, elder statesman, is the inventor of the modern hot air balloon, and international aviation world record holder, which included the first successful crossing of The English Channel by hot air balloon; and*

WHEREAS, *Mr. Yost continues to support and encourage advances in aviation and space flight; and*

WHEREAS, *Mr. Yost continues to inspire and mentor young people with an interest in aviation;*

NOW, THEREFORE I, *Bill Richardson, Governor of the State of New Mexico, do hereby proclaim June 30, 2004 as:*

"ED YOST DAY"

throughout the State of New Mexico, and urge all citizens to realize the importance of aviation and the contributions that Mr. Yost has made to aviation and his country.

Attest:

Done at the Executive Office this 9th Day of June, 2004.

Rebecca Vigil-Giron
Secretary of State

Witness my hand and the Great Seal of the State of New Mexico.

Bill Richardson
Governor

Ed Yost Day proclamation. – *Courtesy State of New Mexico*

Governor Bill Richardson proclaimed June 30, 2004 "Ed Yost Day." The Hot Air Balloon was declared the official air craft of New Mexico by New Mexico lawmakers.

Medallion minted in nickel and bronze was presented to Ed by John Craparo, founder of the International Aeronautic League.

Other honors included the Coupe Chateau de Balleroy, 1977, for contributions to the sport of ballooning, particularly the flight of the *Silver Fox*. Ed was the first living inductee to the International Balloon and Airship *Hall of Fame*.

The Lipton Trophy is awarded to Ed Yost as only the 4th recipient since 1908. – *Photos courtesy of the Journal of the British Balloon & Airship Club*

Ed organized the First World Hot Air Balloon championship, February 11-17, 1973, in Albuquerque, New Mexico, wrote the rules, and served as the Balloon Meister. Once at one of the balloon festivals, a spectator who did not know Ed asked if he had any connection to the extravaganza that was taking place. Ed answered, "Kinda."

Ed designed and built Malcolm Forbes' *Chateau de Balleroy* balloon. He supervised the ground crew and flight operations for Forbes, while Forbes crossed the continental United States.

Ed designed and built the *Eagle, Double Eagle*, and *Double Eagle II* which made the first successful crossing of the Atlantic Ocean by balloon. He also designed and built the *Balloon of Peace* for Joe Kittinger, who made the first solo crossing of the Atlantic Ocean.

Ed was inducted into the Aviation Hall of Fame in South Dakota. When he was offered as a candidate to be inducted into the South Dakota Hall of Fame at Chamberlain, South Dakota, by Keystone resident and historian Bob Hayes, his candidacy was not accepted. Hopefully, this legendary pioneer in South Dakota will one day be recognized and inducted.

Ed's Patents:

2,871,597	Dropping Mechanism
2,924,408	Mechanical Balloon Load Releasing Device
2,937,825	Balloon and Load Bearing Attachment
2,932,469	Balloon System
2,950,882	Balloon Gondola
2,990,147	Balloon Load Attachment Fitting
3,006,584	Balloon Load Lowering Mechanism
3,096,048	Heated Gas Generator for Balloons
3,109,612	Taped Plastic Balloon with Jim A. Winker
3,109,611	Balloon Seam Structure and Method of Sealing Balloon Material
3,116,037	Balloon Body Structure for Towed Balloon
3,128,969	Cartridge Inflated Balloon
3,131,889	Balloon Structure with Release Mechanism
3,112,900	Towed Balloon Lift Control
3,170,658	Rapid Controlled Balloon Inflation Mechanism
3,168,266	Method and Apparatus for Supporting Air-borne Loads

3,229,932	Maneuvering Valve for Hot Air Balloon
3,312,427	Balloon Structure with Launching Cells
3,642,400	Illumination Flare/Balloon "Briteye"
3,670,440	Inflatable Display
4,432,513	Improved Gas-Proof Fastening System for a Non-Rigid Airship

Ed was a fascinating story teller, and told how once he was speaking, and a woman in the audience interrupted him and asked him to stop so she could have a potty break. She didn't want to miss anything. When Amy Ballard from the National Forest Service and I arranged for Ed to speak at one of their traditional Moon Walks, this one out at the Stratobowl rim, the crowd was one of the largest.

Ed Yost either liked or didn't like a person. Fortunately, I was lucky to be classified in the latter. Ed often referred to Pam and myself as his "precious friends." I am humbled and honored to have been a colleague and friend for the past 30 years. Ed would sometimes appear to be grumpy. Once I told Ed I was going to do a balloon program and I asked what he thought about it. He said, "I don't care."

Though Ed was a perfectionist, he would tell me when I was going overboard drawing a specific blueprint: "Don't sweat the small stuff." Ed had a professorial manner. He didn't care for trivia or small talk, especially when projects were going full bore. As he grew older he mellowed with a less brusque demeanor. When we worked on the monuments projects at the Stratobowl he said, "we're involved in a plot put on by Jesus." Then he said, "He's keeping me alive for a reason."

I was talking to Becky Wigland, Curator at the National Balloon Museum in Indianola, Iowa about Ed's donation of a large singer sewing machine used for fabric construction –Ed said, "I call it Kate Smith."

"Why?" "Because it's a big singer."

When Ed's son, Greg, called me in 2007, I immediately interjected and said, "Oh no," for I knew it was a call informing me of Ed's death – May 27, 2007. "Would you conduct Ed's memorial service and committal?" Greg asked. "Of course," I said. There had been a brief celebration of his life at Sipapu, New Mexico June 3rd. So friends and family gathered at the Allison, Iowa VFW. A simple brochure adorned with an elk and a mountain in the background, with the 23rd Psalm printed inside said simply, "In memory of Ed Yost."

My sermon was a Tribute to Ed. Here are a few excerpts:

"Surely the presence of the Lord is in this place
I can feel his mighty power and his grace.
I can hear the brush of angel's wings.
I see glory on each face.
Surely the Lord is in this place."
– Words and Music by Lanny Wolfe. Copyright ©
1977 Lanny Wolfe Music ASCAP. All right controlled
by Gaither Copyright Management. Used by
Permission

"When Joe Kittinger leaped out of that high altitude bal-loon in a parachute over the New Mexico desert over 40 years ago, at 102,800 feet, he prayed this prayer: **'Lord, take care of me now.'**

A good prayer, a worthy prayer – in life and in death.
Ed Yost died a week ago Sunday. He died on the Day of Pentecost – The Festival of the Holy Wind. The New Testament reads: 'The wind blows where it chooses and you hear the sound of it, but you do not know where it comes from or where it goes. So it is with everyone who is born of the spirit.' (John 3:8 — RSV)
Ed was born in Bristow, Iowa–I was born in Bristol, South Dakota. Ed graduated from high school in 1937, the year I was born – the year the Hindenberg exploded and crashed. Neither Ed nor I had anything to do with it... As I drove from Minneapolis this morning, I became aware of 2 additional passengers in my car. I couldn't see them. But I could feel them. I knew they were present there in New Mexico on Sunday as some of you celebrated Ed's life.

I parked my car and these 2 passengers came over—-and just now they are standing nearby. You can't see them. You can feel them.

Their names? ***Loss*** and ***Gratitude.***

Loss is real. Ed is gone. We've lost a giant of a man. Father, grandfather, friend. I expected Ed to live to be 100. So I was stunned. Shocked and shaken when Greg called. But there's another present today. Her name? ***Gratitude!***

Let Gratitude speak:

Thank you Lord, for Ed the Balloonist. Aviator, aeronaut, scientist, inventor, adventurer. Twenty one patents and a lot more. Ed brought magic to the world. The planet has become a garden

with a thousand multi-colored, multi-shaped flowers floating in the air at Albuquerque, Indianola, Sioux Falls, Paris, Africa and Singapore. Just look up. Floating flowers because of Ed Yost.

Thank you Lord, for Ed the dreamer, the visionary

'Imagination,' Einstein said, 'is greater than knowledge.' Imagination plus energy and resources to back it up and dreams happen. Lately, the dream has been an interpretative center at the Stratobowl, sponsored by the Historic Balloon Society. Will the dream live? Will that dream die?

Thank you Lord, for Ed's quirky sense of humor

He always signed his many letters, 'with a smile.'

Once he commented on my letterhead which read Phadness Pharm. He wrote and said 'Phadness Pharm is Phunny.'

Once he wrote a letter from Shadow, his dog, to our 2 dogs, Daisy and Rascal.

Thank you Lord, for Ed the Friend – He either liked a person or he didn't. You and I were lucky to be in the first category. I was often the recipient of his generosity. Always giving me medallions, momentoes, once a load ring from our gas balloon—then this: He gave me a copy of Tom Crouch's book Eagle Aloft. And in the front leaf he wrote – 'June, 2004, to Rev Arley and precious wife

Pam Fadness –the devoted helper of people! Anywhere, Anytime. With admiration – Ed Yost.'

Loss and **Gratitude.**

Rita Mae Brown once wrote, 'I still miss those I love who are no longer with me, but I find I am grateful for having loved them. The gratitude has finally conquered the loss.' This afternoon, I have good news and bad news about Ed Paul Yost. First the bad. It is this: Ed was not perfect. Ed was a sinner like you and me. James 2:10 (RSV) reads, 'Whoever keeps the whole law but fails in one point has become guilty for all of it.' Ed comes to the throne of grace with empty hands, with guilt and sin on his record. But the good news is that our beloved, dear Ed is a sinner of God's redemption in Christ. Redeemed and extraordinarily gifted, Ed was freed to use his gifts for the advancement of humanity and to the glory of God.

Once Ed wrote me in regard to a project we were working on, 'Perseverance, time and prayer will augment our efforts to endure. With a Smile. Ed.'

Today, we commend Ed to the care of Almighty God. We look for the resurrection on the last day. We commend Ed to God with Joe's prayer, 'Lord, take care of me now!'"

Col. Joe Kittinger Jr. said in his own words that "in his career he has received far more than my fair share of awards and honors." Here is a list as compiled from Joe's book, *Come Up and Get Me*, pages 249 and 250. (Reprinted by permission)

Military Decorations

Silver Star with Oak Leaf Cluster
Legion of Merit with Oak Leaf Cluster
Distinguished Flying Cross (Project Manhigh)
Distinguished Flying Cross (Project Excelsior)
Distinguished Flying Cross with four Oak Leaf Clusters (Vietnam)
Bronze Star with "V" device and two Oak Leaf Clusters

Meritorious Service Medal
Air Medal with twenty-three Oak Leaf Clusters
Purple Heart with Oak Leaf Cluster
Presidential Unit Citation
Air Force Outstanding Unit Award
Army of Occupation Medal
National Defense Service Medal
Vietnam Service Medal with seven Service Stars
Republic of Vietnam Cross of Gallantry with Palm
Republic of Vietnam Campaign Medal
Prisoner-of-War Medal

Civilian Decorations

Harmon International Trophy (Aeronaut)
Aeronaut Leo Stevens Parachute Medal
John Jeffries Award, Institute of Aerospace Sciences
Aerospace Primus, Air Research and Development Command
Hall of Fame, United States Air Force Special Operations
Fe'de'ration Ae'ronautique Internationale Montgolfier Diplome
Paul Harris Fellow, Rotary International
Distinguished Achievement Award, Order of Daedalians
Fellow, Society of Experimental Test Pilots
Elder Statesman of Aviation Award, National Aeronautics Association
Barnstormer of the Year, International Society of Aviation Barnstorming Historians
National Aviation Hall of Fame
International Forest of Friendship—Atchison
Wright Brothers Memorial Hall of Fame
Air Force Space and Missile Pioneers Award
Parachute Industry Association
Legion of Merit (Italy)
Santos Dumont Medal, French Aero Club
Le Grande Medaille, City of Paris
Revoredo Trophy; International Flight Research Corporation
Joe W. Kittinger Medal of Achievement, Board of County Commissioners, Orange County, Florida

Heroic Achievement Award, City of Orlando
John Young Award, Orlando Chamber of Commerce
Distinguished Achievement Award, for American former POWs
Achievement Award, Wingfoot Lighter Than Air Society
W. Randolph Lovelace Award, Society of NASA Flight Surgeons
Godfrey L. Cabot Award, Aero Club of New England
Prix de L'Adventure Sportive, French Sporting Adventure Trophy
Award, Chateau de Balleroy
National Air and Space Museum Trophy
(Lifetime Achievement in Aviation Award)
John Young History Maker Award
Florida Aviation Hall of Fame

South Dakota salutes Ed and Joe for bringing adventures and accomplishments that Jules Verne would be amazed and proud of.

Chapter Eighteen

Dakota Ballooning Today

"Still, some means must be found to cross the Atlantic on a boat, unless by balloon...."– Jules Verne, *Around the World in Eighty Days*

The ultimate ballooning prize would remain elusive for a long time. Who would take up the challenge? What daredevil would venture to circumnavigate the earth by balloon? Everyone knew flying the globe by balloon would require great risk, extraordinary courage and sufficient funding. Such a foolhardy endeavor would exclude the faint of heart. It would exempt the aeronaut of modest means.

But then entered millionaire Maxie Anderson and Don Ida. The two aeronauts would make their third attempt to circle the globe by balloon flying out of the Stratobowl of South Dakota. It was November 7, 1982. Maxie and Don launched in the gas balloon dubbed *Jules Verne*. It would be a spectacular adventure. Unfortunately, the envelope developed a fatal leak. The flight was terminated before crossing the Atlantic. They traveled 1162 miles in 16 hours.

Maxie and Don had tried twice before. The first attempt in the *Jules Verne* was from Luxor, Egypt. They landed at Hansa, India, after losing too much precious gas. They had traveled 2,676 miles in 48 hours.

The second attempt also in the *Jules Verne*, was on December 20, 1981, when the two launched near Hansa, India. But because of a serious leak they flew only 20 miles.

The three attempts and failure to circle the globe was a bitter pill to swallow, especially for Maxie who with two other crew members had astonished the world by being the first to cross the Atlantic Ocean by balloon on August 17, 1978.

French writer Jules Verne spurred balloon flight dreams when he wrote the classic *Five Weeks in a Balloon* in 1862. Then in 1873, he wrote *Le Tour du monde en quartre-vingts jours, (Around the World in Eighty Days)*. The story line of *Around the World*, saw Phileas Fogg of London, and his French valet, Jean Passepartout, attempt to circle the globe by rail and steam in 80 days. They were to leave London on October 2, 1872, and were due back December 21, or else they would lose the wager.

Although a leg of the journey by hot air balloon became one of the later images associated with the story – the adventurous symbol of a balloon never actually appeared in Verne's novel.

The idea of a balloon deployed in *Around the World*, is briefly mentioned in chapter 32.

Phileas Fogg dismissed the curious idea of balloon travel when he said, "Still some measure must be found to cross the Atlantic...**unless by balloon** – which would have been adventuresome, besides not being capable of being put into practice." [1]

It took Michael Todd's 1956 movie adaption and triumph of imagination, creating *Around the World in Eighty Days* to dramatize the idea of making balloon flight a part of the mythology of the story.

Captain Joe Kittinger dreamed the "dream to fly around the world," while a prisoner of war in the famous "Hanoi Hilton" during the Vietnam War. "I thought back on my balloon training on the Manhigh and Excelsior and Stargazer days. I began to think about what it would take to make a solo flight around the world in a balloon. It was kind of a romantic, crazy idea- but what a glorious trip it would be! It was one of the last great adventure challenges out there, and I saw no reason why I shouldn't be the one to do it." [2]

Forty years after Michael Todd's four Oscar-winning-movie did the South Dakota Stratobowl become the chosen site, once again, this time for Millionaire Steve Fossett's first attempt to fly around the world. Fossett, a successful business man, launched from the historic bowl in the *Global Challenger,* January 15, 1996. However, trouble plagued the flight soon after takeoff when the mylar skin began to rip free from the envelope. When additional problems developed, Steve Fossett was forced to land near St. John, Newfoundland, Canada, 2 days later. He had traveled 2,200 statute miles.

Among the nearly thirty actual flight attempts, and planned flights that failed to launch for the circumnavigation around the world, was South Dakota's Jacques Soukup. On June 19, 1998, a news conference at the Soukup and Thomas International Balloon and Airship Museum in Mitchell, South Dakota announced "the *Spirit of Peace* around-the-world by balloon flight attempt." The team would consist of Jacques Soukup of Tyndall, South Dakota, Mark Sullivan and Crispin Williams.

The *Spirit of Peace* envelope would be a Roziere balloon, that is, one using a combination of gas and hot air. The balloon was built by Cameron Balloons in the UK. It would hold 550,000 cubic feet of helium gas and 24,000 cubic feet of hot air. When inflated it would tower up to 10 stories high.

The capsule, which would house the aeronauts, would be constructed from Kevlar and carbon fiber. The total rig, planned to be launched from Albuquerque, New Mexico, would seek the jet stream flying at heights up to 42,000 feet.

Spirit of Peace was in a dead heat competition to be the first around the world with Steve Fossett's *Solo Spirit*, and other teams and their balloons, as the *Virgin Challenger, Breitling Orbiter, and Team Re/ Max*. However, due to unfavorable weather predictions, and the success of another competitor, the *Spirit of Peace* never launched. The race was over! History was made at 4:45 A.M., March 20, 1999, when the Breitling Orbiter 3 achieved the first nonstop around-the-world balloon flight. Switzerland's Bertrand Piccard and British Brian Jones had caught the jet stream from the snowy Swiss Alps on March 1st, and landed near Mut, Egpyt, 19 days, 21 hours and 55 minutes later. They traveled over 29,000 miles. The two aeronauts claimed "The Prize and the Glory" so elusive, so long.

But who should reappear on the scene? Steve Fossett, in 2002, would became the first person to fly around the world **solo,** in the *Spirit of Freedom*. The 10-story high balloon was launched from Northam, Western Australia, and returned to Australia in 13 days, 8 hours and 33 minutes. Fossett's balloon was a Roziere outfitted with an auto pilot system and computer controlled burners. The landing was harrowing,

as the balloon envelope dragged Steve along the ground for an ago-
nizing twenty minutes. Steve survived. Another record was claimed!

Recent South Dakota ballooning history has noted a multi-colored
canvas of many significant ballooning places, events, organizations and
people. Here are a few:

In 1989, the world's largest balloon and airship museum was
founded by Aeronauts Jacques Soukup and Kirk Thomas in an unlikely
tiny South Dakota town – Tyndall, South Dakota. Population 600. The
museum exhibited the first balloon basket to fly over the Soviet Union,
rare collections from the Hindenburg Airship, balloon mail, lithographs,
jewelry, trophies and other collectibles. Soukup and Thomas pioneered
specially shaped balloons, three, named *Balloon Hilda, Chesty,* and
Matrioshka. Jacques and Kirk, along with the city of Tyndall, hosted
the 6th World Gas Balloon Championship and the 1st World Roziere
Balloon Championship in 1990 and 1992 respectively.

The Soukup and Thomas International Balloon and Airship Museum
moved to Mitchell, South Dakota near the famous Corn Palace in 1992.
The Mitchell Chamber of Commerce sponsored the "Corn Palace
Balloon Rally." Many notables attended the popular rallies, such as
South Dakota's late Governor Bill Janklow (balloon co-owner with
Sioux Falls pilot Gerald Tennissen), Steve Fossett, Karl Stefan, Mark
West (President of Raven's Aerostar Division), Ed Yost, Don Piccard,
and Mike Wallace. Located just off Interstate 90, the Soukup and
Thomas museum joined South Dakota's pantheon of tourist attractions
and historical sites such as those out west–Mount Rushmore, Crazy
Horse Monument, the Stratobowl Rim Monuments, Badlands National
Park, Wall Drug, Custer State Park, and Jewel and Wind Cave National
Parks. The city of Mitchell provided $650,000 to build the 14,000
square foot museum while paying $150,000 to move the museum's
extensive collection from Tyndall to Mitchell.

However in 2000, the museum closed its doors due to lack of
funding. The city of Mitchell city council had voted 5 to 3 to end
funding to the museum.

The National Balloon Museum of Indianola, Iowa, competed with
the Anderson-Abruzzo Balloon Museum of Albuquerque, to take
over the splendid exhibits. The Albuquerque museum prevailed. The

contents required nine tractor trailer rigs and one flatbed trailer to move the historic cargo to its new and permanent home in New Mexico where it now resides.

"In 1962, Don Piccard, Raven Industries' Marketing Manager, organized the first hot air balloon race. The race was held at the Winter Carnival in White Bear Lake, Minnesota. A sterling silver bowl was given as a prize to give the race credibility." [3]

By 1963, sport ballooning had grown quickly, so to highlight this growth the first annual U. S. Hot Air Balloon Championship took place in Kalamazoo, Michigan. Ed Yost, representing Sioux Falls, flew the 60,000 cubic foot *Channel Champ* balloon he had used earlier that year to cross the English Channel. A reporter wrote humorously and prophetically that hot air balloon races would, "soon outrank the Kentucky Derby and the World Series put together as a stellar sporting attraction." [4]

The Sioux Falls area gave birth to the Sioux Falls Ballooning Association (SFBA), which was formed in 1982. Its mission reads: *The Sioux Falls Ballooning Association is here to provide, by professional means, education, safety and training of both crew and pilots through the promotion of the sport of hot air ballooning in the greater Sioux Empire.*

The Sioux Falls Balloon Association hosts the annual Great Plains Balloon Race, usually held around the second week of June. "There was a one time event, organized and held in 1975, celebrating the Bicentennial. This joint effort between Raven Industries (myself included – Orv. Olivier) and the Sioux Falls Jaycees, was held at Buffalo Ridge, west of Sioux Falls." [5]

The race first started in 1978, and was spearheaded by Orvin Olivier, pilot, and Sales and Marketing Representative for Raven Industries at that time.

Presently, the SFBA has recently built a near exact replica of the first balloon that Ed Yost flew in 1960. Orv. Olivier reported, "we have inflated it a couple of times and are currently making a couple of minor changes. It should be ready for the public unveiling soon." (In 2012 it was displayed and enthusiastically admired at the Albuquerque Balloon Fiesta).

South Dakota hot air balloons – *Photo courtesy Old County Courthouse Museum*

Looking into South Dakota skies today, one sees a bountiful flower garden–floating tulips and roses, petunias and irises, holly hocks and African violets, kissed by the wind.

One can see Orvin Olivier's *Serenity IV and V,* Mark and Kay West's *Scream Catcher, Nimbus 2000, Sunny-Side-Up, Whatever (a balloon with an attitude),* as well as *Cool-in-the-shades* owned with Lyle and Laura Ruesch. *Cloud 9* floats by, piloted by Duane Waack. There's Deven Burnham's *Mystic,* Brad Temeyer's *High Hopes,* and Troy and Carrie William's *Twist and Shout.*

A passenger in a high flying Boeing 787 headed for Reno looks down and sees Stan Burger's red, black and blue *Dream Catcher,* and Dave and Denise Christensen's *Moon Shadow.* Next to Linda Conover's *Checkers* is Vern Feeker's *Great Balls of Fire,* and above those two, Joel Blankers flies *Seventh Heaven.*

In the past we would see Jim Winker's *My Blue Heaven* and Rus Pohl and Bernie Tyrrell's *Elegant Lady,* now both retired. But back in the skies there's the indomitable Jon Kolba, flying *JQuest.* And way out west on a cloudless day, Captain Steve Bauer flies high and serenely over Custer in a black aerostat from his *Black Hills Balloon* fleet. Who could miss the Elks Club annual Balloon Rally in Rapid City?

In 1999, Crooks, South Dakota balloonist, Jon Kolba, decided to try for the world duration record in the AX-6 category. The record at the time stood at 15 hours and 22 minutes. The National Aeronautic Association issued a record sanction to Kolba; it gave him exclusive rights to be the only person attempting the record for a 90-day period.

Mark West, Aeoratar's President and Chief Engineer, had long contemplated developing a double wall balloon for fuel efficiency, and agreed to build an experimental balloon envelope for Kolba's flight. The upper two-thirds of the balloon were constructed with a double wall, creating a layer of insulation to keep in hot air. A single wall construction of silver mylar fabric on the lower third helped reflect heat back into the envelope.

In preparation for his attempt, Kolba spent time in a local meat locker's freezer to experiment with different electrical pieces and his winter outerwear. He tested the equipment and his endurance to stand up to the cold temperatures he would experience during the flight. He also spent many hours preparing for the flight at the local National Weather Service office.

February 12, 2000, Kolba of Crooks, South Dakota set a new duration record in the custom-made double wall balloon designed by Mark West. 6

This new duration record set at Jamestown, North Dakota, added up to an amazing 21 hours and 55 minutes in the air.

In the mid-1990's a Sioux Falls balloon, 2KW, owned by Bill Kullander and Bill Kuhle, once called *Feathers* and used by Jetta Schantz of Florida, set altitude, distance and duration records. [7]

Over the years, many world class balloonists have flown out of South Dakota's Stratobowl. One example is Troy Bradley who is in the record books for at least three flights: Troy Bradley won the 4th American Challenge in 1998 flying in a homemade gas balloon from Albuquerque to Ontario, Canada—or 1,389 miles in 58 hours and 55 minutes. His co-pilot was his wife Tami Stevenson-Bradley. In 1999 Troy flew in the Re/Max Cup winning 6th place. In 2000 he flew in the 5th American Challenge, logging 1,237 miles from Albuquerque to Canada in 52 hours, 25 minutes. Troy's longest record-setting flight was 144 hours and 23 minutes. With Richard Abruzzo, he was the first to fly a balloon from

North America to Africa. Along with Mark Sullivan and Sid Cutter, Troy won the first U. S. National Hot Air Team Championships.

Troy and Tami's son, nine-year-old Bobby, flew a hot air balloon solo near Tome, New Mexico and thereby floated into the history books.

Anyone of any age – nine to ninety- who is yearning to fly South Dakota skies and experience an unforgettable, serene and blissful ride, need only contact one of the several balloon companies presently doing business.

Who are some of these pleasure-givers?

Hot Air Balloon Rides in South Dakota:

Black Hills Balloons
Hot Air Balloon Ride
Call Captain Steve Bauer of Custer
605-673-2520

Cloud 9 Balloons
Sioux Falls, South Dakota
Duane Waack
605-371-1740

Mystic Balloon Rides, Sioux Falls,
South Dakota
Devin and Darcy Burnham

605-940-7706

Praire Sky Hot Air Balloons
Kay West

605-332-5381

Balloon ascension at the Gold Discovery Days in Custer
—*Photo courtesy of Paul Horsted*

EndNotes

Chapter One

1. Paul Fillingham, *The Balloon Book,* (David McKay Co. Inc., New York, 1977), p. 121.
2. Dick Wirth and Jerry Young, *Ballooning,* (Random House, 1980), p. 118.
3. *Ibid.,* p. 118
4. *Ibid.,* Fillingham, p. 121
5. http://en.wikipedia.org/wiki/Francesco_Lana_de_Terzi.
6. *Ibid.*
7. Ward T. Van Orman, *The Wizard of the Winds,* (North Star Press, St. Cloud, Minnesota, 1978), p. 21, 22.
8. Tom D. Crouch, *The Eagle Aloft,* (Smithsonian Institute Press, Washington, D. C., 1983), p. 197

Chapter Two

1. F. Stansbury Haydon, "Letter from E. P. Alexander to A. L. Alexander, September 8, 1861" *Military Ballooning During the Early Civil War* (The John Hopkins University Press, 1968), p. 204.
2. Thaddeus C. Lowe, *Memoirs of Thaddeus S .C. Lowe, Chief of the Aeronautic Corps of the Army of the United States During the Civil War,* (Edwin Mullen Press, 2004).
3. *The New York Times,* June 2, 1862.

4. Jay Monaghan, *The Life of General Armstrong Custer,* (University of Nebraska Press, 1959).
5. George Armstrong Custer, "War Memoirs," *Galaxy, Miscellany and Advertiser,* November 1876, p. 686.6. cited in Crouch.
6. George Armstrong Custer, "War Memoirs," cited in Crouch, p. 386.
7. *Ibid.*, p. 687-688.
8. *Ibid.*
9. Jessie Y. Sundstrom, *Custer County History to 1976,* (The Custer County Historical   Society, 1977).

Chapter Three

1. *Custer Chronicle,* July 1, 1882.
2. *Custer Chronicle,* July 8, 1882.
3. Christine Kalakuka and Brent Stockwell, *Hot Air Balloons,* (Friedman/Fairfax Publishers, 1998), p. 21, 22.
4. *Ibid.*, p. 119.
5. *Eastern Pennington County Memories,* published by The American Legion   Auxillary.
6. Paul E. ("Ed") Yost told this story during a meeting at the Old Courthouse Museum, Sioux Falls, SD, in November, 1985.
7. Virginia Priefert, *Those Who Flew,* (Turner Publishing Company. Paducah, KY., 2002), p. 12.
8. From the Archives at the *Center for Western Studies* at Augustana College, in Sioux Falls, South Dakota.
9. *Ibid.*

Chapter Four

1. Arch Whitehouse, *Heroes of the Sunlit Sky,* (Garden City, New York, Doubleday and Company, Inc.), 1967, p. 94.
2. *Ibid.*, p. 94, 95.
3. Mauer, Mauer, ed., *The U.S. Air Service in World War I* (Washington, D. C., 1978), vol. 1, p. 137.
4. "Thrilling Sport of Ballooning Revived," *Aerial Age,* September 15, 1919, p. 9.
5. *Ibid.*, Mauer, p. 137.

6. *Ibid.,* Whitehouse, p. 90.
7. *Ibid.,* Mauer, p. 138.
8. http://www.militarytimes.com/citations-medals-awards/recipient. php? recipientid=16127.

Chapter Five

1. Interview with Evelyn Chard of Custer, South Dakota, by *Balloons Aloft:*, author, January 7, 2012.
2. *Ibid.,* Crouch, p. 592.
3. *Ibid.,* p. 604.
4. *The Black Hills Engineer Quarterly by the South Dakota School of Mines,* September, 1936, p . 51.
5. *Ibid.,* p. 138.
6. *Ibid.,* p. 139.
7. Tom Walsh, *Stratosphere Bowl* paper delivered at the West River History Conference, April 8, 1978.
8. *Ibid.,* Gregory P. Kennedy, *Touching Space,* (Schiffer Military History, Atgen, Pennsylvania) , p. 25.
9. Albert W. Stevens, "Exploring the Stratosphere" *National Geographic,* October, 1934, Vol. LXVI No. 4, pp. 401-402.
10. Interview with Lola Aimonetto at her home in New Castle, Wyoming, by *Balloons Aloft:*, author, 2005.
11. *Ibid.,* Stevens, pp. 410, 413.
12. *Ibid.,* p. 415
13. David L. Bristow, "Flight of the Explorer" *Nebraska Life,* July/ August, 2007, pp. 56-62.
14. *Ibid.,* Stevens, p. 416.
15. *Ibid.,* Bristow, p. 58.
16. *Ibid.,* p. 61.
17. *Ibid.,* Stevens, p. 417.

Chapter Six

1. Interview with George Moseman by *Balloons Aloft:*, author, in the 1980's .
2. Interview with Luverne Kraemer by *Balloons Aloft:*, author, in 2005.

3. *Stratosphere Bowl* paper by (eyewitness)Tom Walsh , April 8, 1978.
4. *Ibid.*, Walsh paper.
5. Albert Stevens, "Man's Farthest Aloft" *National Geographic Magazine*, January, 1936, pp. 59, 60.
6. Interview with Bob Hayes by *Balloons Aloft:*, author.
7. Interview with Helen Wrede by *Balloons Aloft:*, author, April 21, 2010.
8. *Ibid.*, Kennedy, pp. 24-33.
9. *Ibid.*, p. 31.
10. *Stickney Argus*, Vol 108-No 6, August 20, 2010, p. 12.
11. Paul Higbee,"The Stratobowl Flight" *The South Dakota Magazine*, Nov/Dec. 2005. pp. 37-41.
12. *Ibid.*, George Moseman interview.
13. *Ibid.*, *Stickney Argus*, p. 12.
14. Telephone interview with Robert Plut by *Balloons Aloft:*, author in 2005.
15. *The Aurora County Standard* and *White Lake Wave*, Vol. 128-No. 31 USPS – 03840, August 4, 2010.
16. *The Aurora County Standard* quoted in *The Stickney Argus* Vol. 108 – No. 6 USPS 521-920, August 11, 2010.
17. *The Stickney Argus*, Vol. 108-No. 6 USPS 521-920, August 11, 2010.

Chapter Seven

1. *Ibid.*, Kennedy, p. 43.
2. Interview with Patricia Sanmartin by *Balloons Aloft:*, author, January 12, 2012.
3. *The National Geographic Magazine*, February, 1957, pp. 269-282.
4. *Ibid.*, p. 273.
5. Craig Ryan, *Pre-Astronaut s Manned Ballooning on the Threshold of Space*, (Naval Institute Press, Annapolis, Maryland, 1995), 230.
6. Malcolm Ross, "Plastic Balloons for Planetary Research," *The Journal of the Astronautical Sciences* 5 (Spring, 1958): 5, quoted in Devorkin, "Race to the Stratosphere" p. 307.
7. Navy Historical Center, accessed February 25, 2011, http://history.navy.mil/faqs/faq124-1htm.

8.   Donald Dale Jackson, *The Aeronauts*, (Time-Life Books, Alexandria, Virginia,1980), 133.

Chapter Eight

1.   Gordon Hanson, "Balloonist hope weather eases to allow 50th anniversary flight" *Rapid City Journal*, section C, Sunday, November 10, 1985.
2.   *Ibid.*
3.   *Ibid.*, November 12, 1985.
4.   *Ibid.*
5.   Gilbert Grosvenor, "Forward" *National Geographic Magazine*, March, 1985.
6.   *Ibid.*, *Rapid City Journal*, November 10, 1985.
7.   "Joe Kittinger remembrance," November 1, 2002, quoted by Christine Kalakuka in "Balloon Historical Document," 2004, cited in *Ballooning Magazine*, January/February, 2003.
8.   *Rapid City Daily Journal*, Saturday, November 12, 1960.
9.   *Ibid.*, "Joe Kittinger, Remembrance."
10.   A paper presented at the *Tenth Annual Dakota Historical Conference* in Rapid City, South Dakota.
11.   Briefs, *Rapid City Journal*, 2011.

Chapter Nine

1.   Kevin Woster, reporting, *Rapid City Journal*, July 29, 2004, p. 1.
2.   Robert E. "Bob" Hayes, "Black Hills Hidden History – The Stratobowl" paper delivered at the 39th *Annual Dakota Conference on Northern Plains History, Literature, Art and Archeology,* Augustana College, August 20, 2007.
3.   *Ibid.*, p. 3
4.   *Ibid.*, p. 4
5.   John Cooper "Letter to Arley Fadness," dated August 1, 2006.
6.   Kevin Woster, "Ballooning Pioneer has big dreams" *Rapid City Journal*, p. A4.
7.   *Ibid.*

Chapter Ten

1. Bert Webber. *Silent Siege*, (Ye Galleon Press, Fairfield, Washington), 17.
2. Robert C. Mikesh. *Japan's World War II Balloon Bomb Attacks On North* America "Smithsonian Institution Press, City of Washington, 1973), 2.
3. Lawrence Larsen. *South Dakota State Historical Society and Board of Cultural Preservation Quarterly.* Spring, 1979, Vol. 9, No. 2, p.105
1 Range Supervisor John P. Drissen to Supervisor C.W. Spaulding, 3 Apr., 1945, Decimal Correspondence File 036, Japanese Balloon Report, Cheyenne River Agency, Records of the Bureau of Indian Affairs, Record Group 75, Kansas City Federal Archives and Records Center (hereafter cited as Japanese Balloon Report, RG 75, Kansas City FARC)....
4. *Rapid City Daily Journal*, 20 August, 1945. The pilot and author of the account, published as an exclusive at the end of the war, was Cliff Edwards, the newspaper's managing editor.
1. *5. Ibid., South Dakota History,* p. 106.
2. *6. Ibid.,* Bert Webber. pp. 328-353.
3. *7.* James Raab. *American Daredevil Balloonist, W. H. Donaldson,* (Sunflower University Press. Manhattan, Kansas, 1977), p. 10.
8. Anthony Smith, *Jambo,* (E. P. Dutton & Co., Inc. New York, 1963), p. 30.
9. John McPhee, *Irons in the Fire,* (Farrar, Straus and Giroux: New York), p. 119.
10. http://www.psywarrior.com/RadioFreeEurope.html.
11. http://coldwarradios.blogspot.com/2011/01balloon-flight-for-free-dom-in-usa.html.

Chapter Eleven

1. Joe Kittinger and Craig Ryan, *Come Up and Get Me,* (University of New Mexico Press, 2010), 91.
2. *Ibid.,* p. 94.
3. *Ibid.,* pp. 187, 188.

4. *Ibid.*, p. 54.
5. *Ibid.*, p. 56.
6. *Ibid.*, Gregory P. Kennedy, pp. 111-121.
7. *Ibid.*, Kittinger and Ryan, p. 155.
8. Arley K. Fadness, *Six Spiritual Needs In America Today,* (C.S.S. Publishing Co., Lima, Ohio, 1997).
9. Viktor Frankl, *Man's Search For Meaning,* (Washington Square Press, Published by Pocket Books. New York, London, Toronto, Sydney, Tokyo, Singapore, 1984), pp.93, 95.
10. Max Lucado, *Fearless,* (Thomas Nelson, Nashville, Dallas, Mexico City, Rio De Janeiro, Beijing 2009) 157.
11. *Ibid.*, Kittinger and Ryan, p. 54.
12. Glen Moyer, *Balloon Life magazine,* October, 1990, p. 24.

Chapter Twelve

1. ...http://www.redorbit.com/news/space1112495066/ taking-the-bull-by-the-horns-baumgartner
2. *Ibid.*
3. *Ibid.*
4. Craig Ryan, *Magnificent Failure,* (Smithsonian Books, Washington and London, 2003), 109.
5. *Ibid.*, p. 122.
6. Harl Andersen, AP writer, *Argus Leader,* February 3, 1966.
7. *Ibid.*, Ryan, p. 173
8. http://www.parachutehistory.com/skydive/records/highalt/1934.html
9. *Ibid.*
10. Richard Branson, *Reach For the Skies:Ballooning, Birdmen, and Blasting into Space,* (Penguin, April 28, 2011).
11. *Argus Leader,* February 3, 1966.
12. *Ibid.*, Ryan, p. 179.
13. *Argus Leader,* May 3, 1966.
14. Jim Winker interview, July 5, 2012.

Chapter Thirteen

1. *Raven: Celebrating 50 years of Innovation*. (Raven Industries Inc., 2006), 4.
2. *Ibid.*, Crouch, p. 642.
3. *Ibid.*, p. 642.
4. Bob Ylvisaker, staff writer, *Minneapolis Sunday Tribune*, Sunday, October 2, 1960.
5. *Ibid.*, p. 6.
6. *Ibid.*, p. 6.
7. *Ibid.*, p. 8.
8. Denniston and Moyer, "The Maiden Flight," *Balloon Life*, October, 1990, p. 17
9. *Ibid.*, p. 14.
10. http://www.ballooninghistory.com/raven/ p. 2 of 6.
11. *Ibid.*, p. 12.
12. *Ibid.*, p. 13.
13. *Ibid.*, p. 17.
14. *Ibid.*, p. 22.
15. *Ibid.*, p. 24.
16. *Ibid.*, p. 26.
17. *Ibid.*, pp. 27, 29.

Chapter Fourteen

1. Arthur Jones, *Malcolm Forbes Peripatetic Millionaire*, (Harper and Row Publishers, New York, 1977). 160.
2. http://www.lighterthanair.org/ellis/ed_yost.

Chapter Fifteen

1. "Raven Industries Balloon Display" at the Minnehaha County Old Courthouse Museum, Sioux Falls, South Dakota, 2012.
2. Donald Dale Jackson, *The Aeronauts. Time-Life Books*, (Alexandria, Virginia, 1980), 134.
3. Jan Boesma, *Gordon Bennett Balloon Race*, (The Hague, 1976), cited in Crouch, p. 542.

4. Frank P. Lahm, "The First Annual Aeronautic Cup Race", in *Navigating the Air,* New York, 1907, pp. 34-47; Lahm biographical files, NASM Library, cited in Crouch, p. 544.
5. Lahm, "Cup Race," p. 35, cited in Crouch, p. 545.
6. *Ibid.,* p. 545.
7. *Ibid.,* Van Orman, p. 49.
8. *Ibid.,* p. 50.
9. *Ibid.,* p. 49.
10. Samuel F. Perkins, "Trip of the Dusseldorf," *Aeronautics* December, 1910, pp. 217-218, cited in Crouch, p. 553.
11. 11. *Ibid.,* Van Orman, p. 50.
12. 12. H. Eugene Honeywell, "My Voyage in the International Balloon Races," *Aeronautics* (December 1912) p. 164, cited in Crouch, p. 556.
13. Honeywell, p. 164, cited in Crouch, p. 556.
14. *Ibid.*
15. Undated Clippings, John Berry, biographical file, NASM library, cited in Crouch, p. 556.
16. Crouch, p. 565.
17. *Ibid.,* Van Orman, pp. 97-106.
18. *Ibid.,* p. 105.
19. Craig Ryan, p. 61.
20. Joe Kittinger and Craig Ryan, p. 197.
21. Debbie Fawcett, "Third Time's A Charm for Sweet Rosie O'Grady" *Ballooning Magazine*, summer, 1985, p. 52.
22. http://en.wikipedia.org/wiki/Gordon_Bennett_Cup_(ballooning)
23. *Ibid.,* p. 2.
24. *Ibid.,* p. 2.
25. *Ibid. ,* p. 2.

Chapter Sixteen

1. *Ibid.,* Raab, p. 65
2. *Ibid.,* pp. 68 ff.
3. *Ibid.,* Crouch, p. 658.
4. Glen Moyer, "Ed Yost, Father of the Modern Hot Air Balloon" *Balloon Life*, March, 1987, p. 14.

5.   Ed Yost, "The Longest Manned Balloon Flight" *National Geographic Magazine*, February, 1977, pp. 208-217.
6.   *Ibid.*
7.   Charles McCarry, *Double Eagle,* (Little, Brown and Company –Boston—Toronto), p. 21.
8.   *Ibid.* p. 22
9.   *Ibid.*, Moyer, p. 14.
10. *Ibid.*, Joe Kittinger and Craig Ryan,
11. *Ibid.*, p. 213.
12. *Ibid.*, p. 215.

Chapter Seventeen

1.   Rebecca Svec, reporter, *Hastings Tribune*, May 21, 2001.
2.   Patricia Gabbett Snow, *New Mexico and the West Sunday Journal,* August 11, 2002.
3.   "Hometown Heroes" Christine Kalakuka and Brent Stockwell, *Ballooning Magazine,* Sept/Oct, 2002.
4.   Christine Kalakuka, "Celebrating a Good Life," www.ballonlife.com, 07, 2001.

Chapter Eighteen

1.   Jules Verne, *Around the World in Eighty Days,* (Dell Publishing Co., 1 Dag Hammarskjold Plaza, New York, N. Y., 1964), 224.
2.   Joe Kittinger and Craig Ryan, p. 155.
3.   *Raven Exhibit* at the Old County Courthouse Museum in Sioux Falls, South Dakota.
4.   *Ibid.*
5.   Orv. Olivier correspondence, July 5, 2012.
6.   *Ibid., Raven Exhibit.*
7.   *Ibid.*

Bibliography

Books

Albert, Marvin H. (1965). *The Great Race*. New York: Golden Press.

Alexander, E. P., (September 8, 1861). *Letter To A. L. Alexander.*

Amick, M. L. (1875). *History of Donaldson's Balloon Ascension*. Cincinnati: Cincinnati News Company.

Anderson, Lonzo, (1942). *Bag of Smoke The Story of the First Balloon*. Illustrated by Adrienne Adams. New York: The Viking Press.

Baker, David. (2000). *Inventions from Outer Space*. New York: Random House.

Block, Eugene B. (1966). *Above the Civil War*. California: Howell-North Books.

Boesma, Jan. (1976). *Gordon Bennett Balloon Race.*

Branson, Richard. (2011). *Reach for the Skies: Ballooning, Birdmen, and Blasting into Space*. New York: Penguin.

Brigham, Clarence S. (1932). *Poe's Balloon Hoax*. Metuchen, New Jersey: American Book.

Cameron, Don. (1980). *Ballooning Handbook*. London: Pelham Books.

Cohn, Art, Editor. (1956). Michael Todd's *Around the World in Eighty Days Almanac*. New York: Random House.

Crouch, Tom D. (1983). *The Eagle Aloft Two Centuries of the Balloon in America*. Washington, D. C.: Smithsonian Institution Press.

Domek, Tom and Robert Hayes. (2006). *Images of America Mount Rushmore and Keystone*. Charleston, S. C.: Arcadia Publishing.

Fallen Sons and Daughters of South Dakota in World War II. (2002). Volume 3 of 6. Spearfish High School Project. Spearfish, South Dakota.

Fadness, Arley K. (1997). *Six Spiritual Needs in America Today*. Lima, Ohio: C. S. S. Publishing Company.

Fillingham, Paul (1977). *The Balloon Book*, New York: David McKay Co., Inc.

Frankl, Viktor E. (1984). *Man's Search for Meaning*. Washington Square Press. Published by Pocket Books.
Haydon, F. Stansbury. (1968). *Military Ballooning During the Civil War*. Baltimore and London: The John Hopkins University Press.

Heyerdahl, Thor. (1950). *Kon-Tiki*. London: George, Allen & Unwin.

Jackson, Donald Dale. (1980). *The Aeronauts*. New York: Time-Life Books.

Jarrow, Gail. (2010). *Lincoln's Flying Spies*. Honesdale. PA: Calkins Creek at Boyds Mill Press.

Jaeger, Michael and Carol Lauritzen, Editors. (March, 2004). *Memoirs of Thaddeus C. S. Lowe Chief of the Aeronautic Corps of the Army of the United States During the Civil War: My Balloons in Peace and War. (Studies in American History)*.Lewiston, New York: The Edwin Mellen Press.

Jones, Arthur. (1977). *Malcolm Forbes Peripatetic Millionaire*. New York: Harper and Row.

Kalakuka, Christine and Brent Stockwell. *(1998)*. *Hot Air Balloons*. Hong Kong: Friedman/Fairfax Publishers.

Kennedy, Gregory P. (2007). *Touching Space: The Story of Project Manhigh*. Atglen, Pennsylvania: Schiffer Military History.

Kittinger, Joseph and Martin Caiden. (1961). *The Long and Lonely Leap*. New York: E. P. Dutton and Co. Inc.

Kittinger, Joseph and Craig Ryan. (2010). *Come Up and Get Me*. Albuquerque: New Mexico Press.

Kraemer, Norma J. (2010). *Images of Aviation: South Dakota First Century of Flight*. Charleston, S. C. : Arcadia Publishing.

Lahm, Frank P. (1907). *Navigating the Air*. Cambridge, Massachusetts: Harvard College Library.

Lomax, Judy. (1987). *Women of the Air*. New York: Dodd, Mead and Company.

Lowe, T. S. C. (1899). *Official Report in The War of Rebellion: A compilation of the Official Records of the Union and Confederate Armies*, Series 3, Volume 3. Washington, D.C.: Government Printing Office.

Lucado, Max. (2009). *Fearless*. Thomas Nelson Publishers.

Marriott, Peter and illustrated by John Bavosi and Tom Brittain. (1980). *The Amazing Fact Book of Balloons*. A & P Creative Education. Oakland, California: Worzalla Publishing Company.

Mauer, Mauer, Ed., (1978-79). *The U. S. Air Service in World War I*. 4 Volumes. Washington, D. C.: Government Printing Office.

McCarry, Charles. (1979). *Double Eagle*. Boston: Little Brown and Company.

McPhee, John. (1998). *Irons In The Fire*. New York: Farrar, Straus and Giroux.

Merington, Marguerite, Editor, (1950). *The Custer Story — The Life and Intimate Letters of General George A. Custer and His Wife Elizabeth*. New York: The Devin-Adair Company.

Mikesh, Robert C. (1973). *Japan's World War II Balloon Bomb Attacks on North American*. Washington D. C.: Smithsonian Institution Press.

Monaghan, Jay. (1959). *The Life of General Armstrong Custer*. Lincoln: University of Nebraska Press.

Nelson, Ray. (1985). *Flight of the Pacific Eagle*. Albuquerque, NM: Adobe Press Inc.

Piccard, Auguste. (1956). *Earth, Sky and Sea*. New York: Oxford University Press.

Priefert, Virginia, Compliler. (2002). *Those Who Flew*. Paducah, Kentucky: Turner Publishing Company.

Raab, James. (1977). *American Daredevil Balloonist, W. H. Donaldson*. Manhattan, Kansas.

Raven:Celebrating 50 Years of Innovation. (2006). Sioux Falls, South Dakota: Raven Industries Inc.

Reuss, K. F. (1959). *Jahrbuck der Luftfahrt (Yearbook of Aviation)*. Mannheim, Germany:
Sudwestdeutsche Verlagsanstalt GMBH.

Ryan, Craig. (2003). *Magnificent Failure*. Washington and London: Smithsonian Books.

Ryan, Craig. (1995). *The Pre-Astronauts*. Annapolis, Maryland: Naval Institute Press.

Simons, David G. (1960). *ManHigh*. Garden City: New York: Doubleday and Company.

Smith, Anthony. (1963). *Jambo African Balloon Safari*. New York: E. P. Dutton and Company Inc.

Stansbury, Haydon F. (1941). *Aeronautics in the Union and Confederate Armies, with a Survey of Military Aeronautics Prior to 1861*. Vol. 1, Baltimore: John Hopkins Press.

Stehling, Kurt R. *Bags Up!* (1975). Chicago, Illinois: Playboy Press.

Steven, Ruth, (1987). *My Husband the First Astronaut*.

Sundstrom, Jessie Y. (1977). *Custer County History to 1976*. Custer, South Dakota: Custer County Historical Society.

Upson, Ralph H. and Charles deForest Chandler. (1926). *Free and Captive Balloons*. New York: The Ronald Press Company.

Van Orman, Ward T. (1978). *The Wizard of the Winds*. St. Cloud, Minnesota: North Star Press.

Verne, Jules, (1981). *Around the World in Eighty Days*. New York: Laurel-Leaf Books, Dell Publishing Co., Inc.

Verne, Jules, (1863). *Cinq Semains en balon (Five Weeks in a Balloon)*. Paris: J. Hetzel Publisher.

Verne, Jules, (1874). *L'll Mysterieuse (The Myterious Isand)*. Paris: J. Hetzel, Publisher.

Waligunda, Bob and Larry Sheehan. (1973). *The Great American Balloon Book*. New Jersey: Prentice-Hall Inc.

Webber, Burt. (1983). *Silent Siege Japanese Attacks Against North America in World War II*. Fairfield, Washington: Ye Galleon Press.

Whitehouse, Arch. (1967). *Heroes of the Sunlit Sky*. Garden City, New York: Doubleday and Company Inc.

Winters, Nancy. (1997). *Man Flies: The Story of Alberto Santos-Dumont Master of the Balloon*. Hopewell, New Jersey: Ecco Press.

Wirth, Dick and Jerry Young, (1980) *Ballooning The Complete Guide to Riding the Winds*. New York: Random House.

Periodicals

Abruzzo, Ben L. "Across the Pacific by Balloon." *National Geographic Volume 161. No. 4*: 1982.

Abruzzo, Ben L. "Double Eagle II: Has Landed." *National Geographic*: December, 1978.

Anderson, Maxie and Kristian. "Balloon 'Kitty Hawk' Crosses North America." *National Geographic*, Volume 158. No. 2: 1980.

Andrews, John. "The Fassbender Family Photographers." *South Dakota Magazine*, May/June 2011: 52-59.

Bristow, David L. "Flight of the Explorer." *Nebraska Life*, July/August 2007: 56-62.

Burnett, David. "A Wild and Ill-fated Balloon Race." *National Geographic*, Volume 164, No. 6: 1983.

Custer, George Armstrong. "War Memoirs." *Galaxy, Miscellany and Advertiser*, November 1876. (A magazine of Entertaining Reading). Vol. 21, January 1876 to June, 1876.

Denniston, George and Glen Moyer. "The Maiden Flight." *Balloon Life*, October, 1990: 12-17.

Fawcett, Debbie. "Third Time's a Charm for Rosie O'Grady." *Ballooning The Journal of the Balloon Federation of America*, Summer, 1985: 52-54

Grosvenor, Gilbert. "The Flight of Rodney the Jazz Bird." *National Geographic*, Volume 169. No 4: 1986.

Hamilton, Tom. "Palm Springs Gordon Bennett Balloon Race." *Balloon Life*, September, 1987.

Higbee, Paul. "The Strato bowl Flight." *South Dakota Magazine*, November/December, 2005.

Hunhoff, Bernie. "Ed Yost Flies South Dakota." *South Dakota Magazine*, May 1987: 6-11.

Kittinger, Joseph. "Transatlantic Solo by Balloon." *National Geographic*, Volume 167, No. 2:   1985.

Larsen, Laurence. "War Balloons Over the Prairie: Japanese Invasion of South Dakota."
South Dakota History Quarterly Spring, 1979: 103-115.

Ludwig, Ruth. "A Visit with the Yosts." *Ballooning: The Journal of the Ballooning Federation of America*, Spring, 1994: 4-7.

McCarry, Charles. *National Geographic's 100 Years*, Volume 174, No. 3, 1988.

Moyer, Glen. "Ed Yost Father of the Modern Hot Air Balloon." *Balloon Life*, March, 1987.

National Geographic Society Collector's Edition Volume 10, *Exploration*, 2005.

National Geographic Society – U.S. Army Corps. "Stratosphere Flight of 1934 in the Balloon Explorer." *National Geographic Society Technical Papers,* 1935.

National Geographic Society – U.S. Army Corps. "Stratosphere Flight of 1935 in the Balloon Explorer II." *National Geographic Society Technical Papers,* 1936.

"NEWS" *Ballooning The Journal of the Ballooning Federation of America Summer,* 1986: 33-40.

Piccard, Bertrand. "Around at Last." *National Geographic,* Volume 196. No. 3, 1999.

Pomerantz, Martin A. "Trailing Cosmic Rays in Canada's North." *National Geographic,* Volume 111, No.1, 1953.

Read, R. and D. Rambow. "Hydrogen and Smoke: A Survey of Lighter-Than-Air Flight in South Dakota prior to World War I." *South Dakota State Historical Society Quarterly,* Volume 18, No. 3: 132-151.

Ross, Malcolm. "Project Strato Lab #5: We saw the World from the Edge of Space."
National Geographic, November, 1961.

Ross, Malcolm. Lewis Lee. "To 76,000 Feet By Strato Lab Balloon." *National Geographic,* February, 1957.

Simpich, Frederick. "South Dakota Keeps it's West Wild." *National Geographic Magazine,* Volume 174, No. 3, 1988.

Stevens, Albert W. "Exploring the Stratosphere." *National Geographic Magazine,* Vol.LXVI, No. 4. October, 1934.

Stevens, Albert W. "Man's Fartherest Aloft." *National Geographic Magazine,* January 19, 1936. pp. 59,60.

Stevens, Albert W. "Scientific Results of the Stratosphere Flight." *National Geographic*, Volume LXIX, No. 5, 1936.

The Black Hills Engineer Quarterly. Published by the South Dakota School of Mines
Rapid City, South Dakota, 1936.

"The Thrilling Sport of Ballooning Revived." *Aerial Age*. September 15, 1919. New York.

The War of the Rebellion: A Compilation of the Official Records of the Union and Confederate Armies. *U. S. War Department,* Series 1, Vols. 5, 10, 11 and 25; Series 2, Vol. 2; Series 3, Vol. 3. Washington, D.C: Government Printing Office, 1880-1901.

Yost, Ed. "Longest Manned Balloon Flight." *National Geographic,* Volume 151, No. 2, 1977.

Newspapers

Hanson, Gordon. "Balloonists hope weather eases to allow 50th anniversary flight." *Rapid City Journal,* August 10, 1985.

_____*Hastings Tribune*. Hastings, Nebraska, May 21, 2001.

Herrick, Howard. "Story of the Stratosphere Balloon Memorial Constructed 14 Miles Southwest of White Lake. A 50 Year Old Dream Finally Realized." (Reprinted from the Aurora County Standard, November 28, 1985). *Stickney Argus,* August 11, 2010.

_____*Minneapolis Sunday Times,* Sunday, October 2, 1960.

_____*New Mexico and the West Sunday Journal,* August 11, 2002.

"New Stratosphere Balloon Monument Planned By Citizens Committee." *Aurora County*

Standard and *While Lake Wave,*Volume 128-No. 33, August 4, 2010.

"Piantanida." *Argus Leader*, May 3, 1966.

"75th Anniversary Stratosphere Balloon Celebration." *Stickney Argus,* Volume 108-No. 6,
August 11, 2010.

"Stevens and Anderson Up Near 14 Miles in Strato." *The Rapid City Daily Journal*, No. 16397. November 16, 1935.

"The Balon Ascension." *Custer County Chronicle*. July, 1882.

_____*The New York Times*. June 2, 1862.

"Tenth Anniversary Celebration." *Rapid City Journal*, August 20, 1945.

"This Weekend Will Mark 75th Anniversary of Historic Stratosphere Balloon Landing."
Aurora County Standard and While Lake Wave, Volume 128 – No. 31. August 16, 1935.

"We Were There....Accounts of the Stratosphere Balloon Landing (From Those Who Saw it in 1935). *Aurora County Standard and White Lake Wave*, Volume 128-No. 34, August 25, 2010.

Web Sites

Commings, Richard H. (January 13, 2011). *Cold War Radios*. Retrieved from http://coldwarradios.blogsopt.com/2011/01balloon-flight-for-freedom-in-usa.html

Friedman, Herbert A. (n.d.). *Free Europe Press Cold War Leaflets*. Retrieved January 3, 2013, from http://www.psywarrior.com/RadioFreeEurope.html
High Altitude World Record Jumps (n.d.). Retrieved March 16, 2012 from http://www.parachutehistorycom/skydive/records/highalt/1934.html

Malcolm David Ross, Balloonist (n.d.). Retrieved January 3, 2013, from http://history.navy.mil/faqs/faq124-html

MilitaryTimes A Gannett Company (2013) *Hall of Valor.* Retrieved January 1, 2013 from http://www.militarytimes.com/citations-medals-awards/recipient.php?recipientid=16127

Raven Industries (n.d.). *The Early Years of Sport Ballooning.* Retrieved January 3, 2013 from http://www.ballooninghistory.com/raven

Taking the Bull By The Horns: Baumgartner completes 132 mile Skydive (March 16, 2012). Retrieved March 15, 2012, from http:www.redorbit.com/news/space1112495066/taking-the-bull-by-the-horns-*baumgartner...*

The Forbes Transcontinental Balloon (n.d.). Retrieved January 3, 2013 from http://www.lighyerthanair.org/ellis/ed_yost.html

Wikipedia The Free Encyclopedia(November 19, 2012). *Francesco Lana de Terzi,* Retrieved January 3, 2013, from http://en.wikipedia.org/wiki/Francesco_Lana_de_Terzi

Wikipedia, The free encyclopedia (December 4, 2012). *Gordon Bennett Cup (Ballooning).* Retrieved January 3, 2013, from http://en.wikipedia.org/wiki/Gordon_Bennett_Cup_ballooning

CPSIA information can be obtained at www.ICGtesting.com
Printed in the USA
LVOW120004210613

339566LV00002B/2/P